Advanced praise for

ENGINEERING DEVOPS

"Enterprises are brushing off the sleep of being industrial companies and awaking as software and analytics companies utilizing Agile and DevOps practices become central to their success. In the fray of books operationalizing DevOps, *Engineering DevOps* stands apart. It alone provides valuable engineering blueprints and a comprehensive collection of engineering practices. The seven-step engineering transformation blueprint is a helpful new approach for organizations needing to transform their delivery pipelines and processes. Marc's Nine Pillars approach to describing DevOps is easily consumable, describing key elements that enable teams to begin their DevOps approach regardless of their current maturity. As a developer-turned-marketeer, I applaud this book for its message and guidance to the industry, and useful guidebook in the journey to improve software value streams."

—**Jeff Keyes**, Product Marketing, Plutora

"Before you hire your next DevOps consultant, I strongly recommend you first get a copy of *Engineering DevOps* by Marc Hornbeek. This is one of most impactful, well-articulated DevOps management toolkits you wished you had years ago. Marc's multidimensional models of DevOps really help organizations understand how to engineer and scale a truly agile enterprise."

—**Patrick Johnson**, General Manager of
Automation Platform Technologies,
Spirent Communications

"As an engineer by training myself, I am totally impressed by Marc's structured engineering mindset in action throughout this book. Starting from a DevOps 'Reference Blueprint' to a complete step-by-step guide to implementing the blueprint for your specific needs and organizational structure, *Engineering DevOps* is truly an Engineering Reference Guide. I will be adding this book to my personal collection of DevOps 'must-reads' and referring it to my clients and peers. Whether you are an engineer yourself, a DevOps implementation leader scaling DevOps across the enterprise, or someone trying to just get started with DevOps, this book has guidance and tips for you."

—**Sanjeev Sharma**, Cloud and DevOps Transformation and Strategy Executive, Thought Leader, Startup Advisor, Keynote Speaker, and Author

"DevOps is a complex maze that has many leaders frustrated. In many large enterprises, 'DevOps' is simply a buzzword they are striving to achieve but struggle with what the end state looks like or even where to start. There are many layers across the organization that are instrumental to truly executing on DevOps. DevOps is not something you go get a quote for and simply buy. It's an evolving journey. In his book *Engineering DevOps*, Hornbeek takes you on the journey in a prescriptive way that leaders and practitioners can execute upon. I've seen Hornbeek lead F500 Enterprises on this multi-year journey firsthand. Hornbeek takes a complex web and breaks it down into an executable approach. He is adamant that engineering is critical on this journey, and this book drives the focus from that perspective. Marc's insights captured here will be of incredible value to you on your personal DevOps journey."

—**Ryan Harmon**, Strategic Account Executive, Trace3

"*Engineering DevOps* is a fantastic guide combining strategy and engineering. It provides valuable tools and practices that every DevOps practitioner needs. In this book, Marc draws up the blueprints to start, improve, and scale DevOps in any organization. I highly recommend it."

—**Aymen El Amri**, Founder, Faun.dev and Eralabs

"When I first founded DevOps.com six-plus years ago, there were a lot of people talking about DevOps but not a lot of people who really understood DevOps. Marc Hornbeek was one of the people who understood DevOps. Marc understood DevOps from many different viewpoints as well, from a long, distinguished career in test engineering to managing a team and all the way to how DevOps could best be implemented across an entire enterprise. In his book, *Engineering DevOps*, Marc brings all of this together. Whether you are a single engineer looking to further your career with some DevOps knowledge or a C-level executive seeking to make your organization more competitive via digital transformation, this is a must-read book for you."

—**Alan Shimel**, Founder, Editor-in-Chief, DevOps.com

"An engineering mindset must carry through every aspect of software delivery from ideation to value creation to compete in a digital landscape. This book by Marc Hornbeek, a well-known and award-winning IEEE engineer, provides solid guidance on adopting DevOps with an engineering perspective."

—**Jayne Groll**, CEO, The DevOps Institute

"Marc has the unique ability to take complex DevOps concepts and distill them down to a language that the business can understand, which is so valuable in a space that has yet to be fully adopted. This book is not just a good read, it is a *must*-read if you are in enterprise IT and even a savvy line of business that understands the power that DevOps brings to product development."

—**Sandy Salty**, CMO, Trace3

"*Engineering DevOps* is a super valuable resource for all those who are looking to optimize their technology strategy. Yes, DevOps is for you! DevOps is more than a rigid definition or specific set of steps. This book does a great job of establishing a framework and understanding that one can extract and apply to their unique variables and requirements."

—**Dan Guzman**, Account Executive, Salesforce

"At the 2016 Institute of Electrical and Electronics Engineers (IEEE) Extraordinary Lives event in 2016, Marc Hornbeek's understanding of what makes automation successful was centerstage. This was an evening hosted by the IEEE Buenaventura Section and a stage to honor someone who had significantly 'advanced technology for humanity.' I am so glad that Marc's DevOps expertise is captured in print, because the words that Marc spoke at the event expressed a wisdom gained through extraordinary contributions to the field. The book *Engineering DevOps* is the answer to the many calls for an independent reference that transcends applications and industries and describes under the same umbrella both strategy and best practice. I want to thank Marc for the commitment and time to coalesce his knowledge in engineering DevOps. The pages vibrate with the same generosity and clarity as when Marc explains the DevOps recipe to an audience or in one-on-one conversation. This blueprint for successful DevOps is a catalyst for industry practice changes and will become a classic on leaders' and practitioners' e-bookshelves."

—**Nathalie Gosset**, BSEE, MSEE, MScTel, MBA, PMI-PBA, Buenaventura IEEE Past Chair, and ISEE Principal Consultant

"DevOps is not just a portmanteau used to scare Development and Operations IT professionals. It's not just marketing hype, and it's not a tactical plan. DevOps is a critical approach to deal with the digital world. *Engineering DevOps* is a comprehensive collection of strategic planning tools, engineering blueprints, and engineering practices that IT leaders and practitioners can use to specify successful implementation plans. It approaches the subject from an engineering practitioner's perspective while tying back to the underlying philosophies. The book is written in such a way as to be useful regardless of your current maturity level. However, the more you know about DevOps, the more you will get out of it."

—**Tim Coats**, Director of Applied Innovation, Trace3

"This book is going to be a prized possession for everyone in IT Industry. It is a must-read for everyone who is planning, implementing, operating, and scaling DevOps. It covers every aspect of DevOps, including how to start an initiative, assess the current state, how to get the team in place, tools, CI/CD pipeline, maintenance, and how to implement continuous improvements throughout the journey of maturing development, operations and Infrastructure. The templates, tips, and tools are useful for practitioners. It is the most comprehensive book to start and run a DevOps initiative."

—**Niladri Choudhuri**, Founder and CEO,
Xellentro Consulting Services LLP and DevOps India Summit

"A lot of IT professionals are exposed to the DevOps ethos via white papers, reading a book, at a conference or trade show, or maybe even anecdotally from business colleagues and contacts...but where do you start? How do you improve? Where are you in your DevOps journey? Marc Hornbeek, a DevOps expert who I've had the privilege of working directly with on several DevOps projects, provides the blueprint for your DevOps journey in his book *Engineering DevOps*, no matter where you find yourself in the DevOps spectrum. If there is one DevOps book you should own, it should be this one."

—**David Vieler**, Senior Director of Operations,
Trace3 Northern California

"In my years of learning, practicing, and teaching DevOps, the lack of a singular guide to the path of DevOps was a frequent problem. They were all incomplete in either scope or detail, and ultimately left me wanting. *Engineering DevOps* is the first book I've seen that does the topic justice in scope and still provides enough detail to be an actionable guide. It doesn't fall into the trap of approaching DevOps like a single discipline, and instead it rightly treats it as the synthesis of a multitude of skills, focused through the lens of rigorous DevOps philosophy and engineering practices. In these pages, you can see what happens when DevOps grows up."

—**Quentin Hartman**, Director of Engineering, Finalze

"I wish I had *Engineering DevOps* to guide me when I started working in SCM and build/release engineering 20 years ago! It clears up many confusing DevOps concepts and myths. It gives real-world examples that can be easily applied and a clear recipe on how to best practice DevOps for the whole organization. My CI/CD pipeline designs follow the 'Continuous Delivery CI/CD Pipeline' blueprint."

—**Somboon Ongkamongkol**, DevOps engineer,
Spirent Communications

"Working in an industry that is yet to settle on an accepted 'definition' for DevOps makes each day a challenge. The ever-morphing landscape makes it difficult for technology vendors and consumers alike to ensure we're all talking about the same thing in the same way. Enter Marc Hornbeek, who has been one of the industry veterans I count on to lend clarity and a direction to this maturing space. In writing *Engineering DevOps*, Marc has given us all some common ground to lend some consistency to our discussions and our best practices. Thank you, Marc!"

—**Doug Miller**, Senior Director Strategic Partnerships,
Perforce Software

ENGINEERING DEVOPS

ENGINEERING DEVOPS

From Chaos to Continuous Improvement... ***and Beyond***

A New Engineering Blueprint for DevOps Transformations

Marc Hornbeek

a.k.a. DevOps_The_Gray esq.

Cover illustration by Dennis Currie
Cover and book design by Devon Smith

ISBN: 978-1-54398-961-8

ENGINEERING DEVOPS

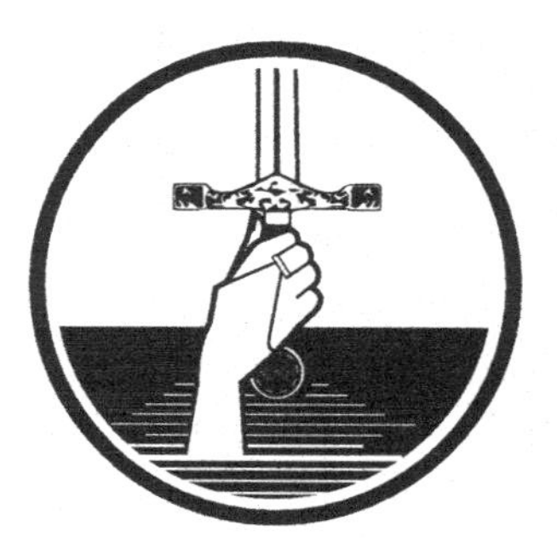

I dedicate this book to my parents, Bill and Jeanette, who always encouraged me to work hard and study; to my children, Sarah-Jane, Michael, and Mathew, and my grandson, Jasper, who are my life's pride; and to my sister, Christine, who has been a steady rock throughout my life.

My mother gave me the following poem as a little boy, and it has inspired me throughout my life.

The heights by great men reached and kept were not
attained by sudden flight, but they, while their
companions slept, were toiling upwards in the night.

—Henry Wadsworth Longfellow

Acknowledgments

This book represents information learned, accumulated and practiced over my engineering career spanning more than forty-three years. Much of my career work has centered on leading engineers, automating and optimizing engineering processes for development, testing, and releases of software systems. While unintended, much of the time the experiences and the knowledge from these projects became essential to a deep and broad understanding of DevOps and lean processes. DevOps emerged as an "important thing" (yet not well defined!) in the IT industry in the later part of my career. I am grateful to my reviewers, sponsors, many colleagues, customers, family, and friends who contributed directly or indirectly to the creation of this book.

Special thanks go to the sponsors of this book who have helped shape the content, offset the cost of production, and will be helping to market this book:

Plutora is a complete value-stream management (VSM) platform that improves the speed and quality of complex application delivery and provides complete visibility of the entire process across the enterprise portfolio (https://www.plutora.com/). **Jeff Keyes, Director of Marketing at Plutora**, is a persistent collaborator for VSM applications to DevOps and a strong supporter of my work and this book.

Spirent Communications brings clarity to increasingly complex technology and business challenges. They offer active service assurance, cybersecurity, high-speed networking, positioning, and navigation, as well as 5G testing, and network virtualization (https://www.spirent.com/). **Patrick Johnson, General Manager Automation Platform Technologies (APT) at Spirent Communication**, is a long-term mentor and strong supporter of this book. We collaborated on content for DevOps applied to network manufacturers, operators, and NetDevOps.

Special thanks go to **Gary Gruver** (www.garygruver.com), a well-known DevOps evangelist and author who generously shared his experience writing DevOps books and provided critical guidance regarding the process of book creation and preparing the book for self-publishing.

Special thanks go to the talented book editorial and publishing team who have shaped the content, look, and feel of this book:

- **Kate Sage**, editor of this book and the editor of *The Phoenix Project*, *The DevOps Handbook*, and other IT Revolution publications. Kate's advice is priceless.
- **Devon Smith**, book designer.
- **Sarah Currin**, proofreader and a proofreader for some IT Revolution publications.
- **Dennis Currie**, cover artist and husband to my niece Lori White.
- **Christina Ramirez**, BookBaby self-publisher.

Extra special thanks go to my mentors and informal reviewers, who have given their personal time, expertise, encouragements, and advice to make this book better than I could have made it without them:

- **Alan Shimel**, Founder and CEO of DevOps.com and personal mentor.
- **Jayne Groll**, CEO of *The DevOps Institute* and personal mentor.
- **Sandy Salty**, CMO of Trace3 and personal mentor
- **Christine White**, my sister, advisor, and chief encouragement officer.
- **Michael Hornbeek**, my son, provided inspiration and suggestions to this book more fun.

- **Jamal Ketabi**, long-time colleague, friend, and Engineering VP.
- **Niladri Choudhuri**, founder and CEO of Xellentro Consulting, India, and DevOps collaborator.
- **Joan Lucas**, personal advisor and friend.

I would like to thank the colleagues who most influenced my work over the years, including those from Bell-Northern Research, DevOps Institute, DevOps.com, IEEE, Trace3, ECI Telematics, GSI Lumonics, Pepperdine University, National Research Council of Canada (NRC), International Standards Organization (ISO), National Institute of Standards and Technology (NIST), Queen's University at Kingston, Spirent Communications, Tekelec, Communications, University of Ottawa, and Xellentro.

I would like to thank the customers with whom I had the opportunity to interact with on various DevOps projects over the years: AAA, AIG, Alcatel, Apple, Bay Networks, Bell Canada, Bell-Communications Research, Blackhawk Networks, BNR, Capital Group, Charter, Cisco, Century Link, Comcast, DaVita, Deutsche Bundespost, DoCoMo, Ericsson, Federal Reserve Bank, France Telecom, GE Digital, IBM, Herbalife, Juniper Networks, LDS, MCI, Microsoft, NetApp, Nokia, NorTel, NTT, PacBell, PimCo, Prince William County, Ross, SGB, Sprint, Spirent Communications, Tektronix, and Telefonica.

Gurus, authors, and evangelists that most influenced my DevOps work include **W. Edwards Deming**, known primarily for his work on total quality management; **Gene Kim**, a DevOps evangelist and co-author of *The Phoenix Project*; **Jez Humble**, co-author of *Continuous Delivery*; **John Willis**, co-author of *The DevOps Handbook*, **Eliyahu M. Goldratt**, author of *The Goal*; **Alan Shimel**, founder of Devops.com; **Jayne Groll**, CEO of The DevOps Institute; **Glenford Myers**, author of *The Art of Software Testing*; **Eberhard Wolff**, author of Microservices; **Nicole Forsgren, Ph.D.**, co-author of *Accelerate*; **Sanjeev Sharma**, author of *The DevOps Adoption Playbook*; and **Betsy Byer**, co-author of *Site Reliability Engineering.*

Product vendors and partners that directly influenced my understanding of DevOps tools products include those from AppDynamics, Atlassian, AWS, Azure, CloudBees/Electric Cloud, Datical, DBMaestro, Docker, Chef,

Cisco, Coverity, Cloudhedge, DBMaestro, Dynatrace, GitHub, Google, Harness, HashiCorp, Fugue, Klocwork, Lenovo, Microsoft, Moogsoft, NetApp, New Relic, OpexSoftware, OverOps, Perforce Software, Plutora, Puppet, Quali, RedHat, Service Now, Spirent Communications, Splunk, Trace3, Tricentis, Veracode, and XebiaLabs.

I am very grateful for close family and friends that supported me and provided motivation during my working long hours in pursuit of an engineering and DevOps, including **Linda Hornbeek**, **Sarah-Jane Morin**, **Michael Hornbeek**, **Mathew Hornbeek**, **Christine White**, **Jamal Ketabi**, and **Somboon Ongkamongkol**.

—Marc Hornbeek
Principal Consultant—DevOps
a.k.a. DevOps_the_Gray Esq.
www.EngineeringDevOps.com
https://linkedin.com/in/marchornbeek

Le Morte d'Arthur— A DevOps Engineering Journey

The stories of King Arthur have fascinated me from the time I was young. Imagine a boy raised in Britain by the wizard Merlin who becomes king because he can pull the magic sword Excalibur from a stone. Once king, he unites warring tribes, forms a brotherhood of knights, and creates a round table to be sure everyone who sits around it has equal status as they pursue the quest for the Grail. I often reflect on these stories and relate them to experiences in my life—including DevOps engineering!

DevOps is the Grail of software creation and delivery. To engineer DevOps, even *The First Way* of DevOps—*Continuous Flow* (harmony)—requires faithful employment of leadership (King Arthur), collaborative team culture (Knights of the Round Table), disciplined engineering process (Laws of Camelot), technology (Excalibur), and automation (the magic of Merlin, Archimedes, and Nimue).

The ultimate pursuit of *The Third Way* of DevOps—*Continuous Improvement* (Quest for the Lost Grail)—requires persistent commitment to engineering discipline and noble practices.

DevOps also requires a persistent evangelist (Nimue, the Lady of Avalon) and leadership (Morte d'Arthur) to be forever vigilant and ready to lead DevOps' evolutions into the future (King Arthur is forever ready to rise again with Excalibur). Both King Arthur and

DevOps depend on the virtuous tenets of chivalry/culture, laws/processes, and use of Excalibur/technology.

Arthurian tales represent a masterful integration of historical events with misty imagined magical episodes and, like DevOps, embody tenets of chivalry and respectful moral leadership. While the means can be elusive, the results can be magical. The sword Excalibur and the search for the Grail are powerful symbols that remind me to always try to engineer technology for good and noble purposes. The stories of King Arthur teach us to lead passionately when needed to triumph against challenges and to continue the pursuit of perfection.

Each of the major parts of this book start with a quote from Arthurian literature and provide my own interpretation from a DevOps point of view. I hope you find this Arthurian analogy as relevant and interesting as I do, and I hope that it adds a little uniqueness and fun to your reading adventure. Now on with the quest of engineering DevOps!

Contents

List of Figures xxi
Introduction xxv

PART I: WHAT IS ENGINEERING DEVOPS, AND WHY IS IT IMPORTANT?

1. What is Engineering DevOps? 3
2. Nine Pillars of Engineering DevOps 17
3. Why is Engineering DevOps Important? 29

PART II: ENGINEERING PEOPLE, PROCESS, AND TECHNOLOGIES FOR DEVOPS

4. How Should People, Process and Technology be *Engineered* for DevOps? 45
5. Value Stream Management (VSM) 71
6. Application Release Automation (ARA) 77
7. Version Management 83
8. Continuous Security (a.k.a. DevSecOps) 89
9. Service Catalogs Facilitate DevOps Engineering 99
10. DevOps Governance Engineering 107
11. Site Reliability Engineering (SRE) 115
12. DevOps Disaster Mitigation and Recovery 121

PART III: ENGINEERING APPLICATIONS, PIPELINES, AND INFRASTRUCTURES FOR DEVOPS

13. DevOps Application Engineering 125
14. CI/CD Pipeline Engineering 145
15. DevOps Elastic Infrastructures 161

16. Continuous Test Engineering 195
17. Continuous Monitoring Engineering 203
18. Continuous Delivery and Deployment Engineering 211

PART IV: DEVOPS SEVEN-STEP TRANSFORMATION ENGINEERING BLUEPRINT

19. DevOps Seven-Step Transformation Engineering Blueprint 225
20. Step One: Visioning 229
21. Step Two: Alignment 233
22. Step Three: Assessment 239
23. Step Four: Solution 245
24. Step Five: Realize 253
25. Step Six: Operationalize 267
26. Step Seven: Expansion 271
27. Future of Engineering DevOps—Beyond *Continuous Improvement* 279
28. Continuous Learning 285

PART V: APPENDIX AND REFERENCES

Appendix A Definition of DevOps Engineering Terms 295
Appendix B DevOps Transformation Application Scorecard 305
Appendix C DevOps Transformation Vision Meeting 307
Appendix D DevOps Transformation Goals Scorecard 311
Appendix E DevOps Transformation Alignment Meeting 313
Appendix F DevOps Transformation Practices Topics Scorecard 319
Appendix G DevOps Assessment Discovery Survey 321
Appendix H DevOps Transformation Practices Maturity Assessment Workshop 327
Appendix I DevOps Current State Value-Stream Mapping Workshop 331
Appendix J DevOps Solution Requirements Alignment Matrix 335

Appendix K Value-Stream Map Template 337
Appendix L DevOps Tools and Comparison Charts 339
Appendix M Engineering DevOps Transformation RoadMap Template 343
Appendix N Engineering DevOps Transformation Backlog Template 345
Appendix O Engineering DevOps Transformation ROI Calculator 347
Appendix P DevOps Transformation Solution Recommendation Meeting 349
Appendix Q NetDevOps Blueprint 357
References 361
About the Author 367

Figures

PART I: WHAT IS ENGINEERING DEVOPS, AND WHY IS IT IMPORTANT?

Figure 1 DevOps Engineering Blueprint 4
Figure 2 CALMS Model (credited to Jez Humble, co-author of *The DevOps Handbook*) 7
Figure 3 Nine Pillars of Engineering DevOps 19

PART II: ENGINEERING PEOPLE, PROCESS, AND TECHNOLOGIES FOR DEVOPS

Figure 4 Capability Maturity Model 50
Figure 5 Engineering DevOps Maturity Levels 52
Figure 6 Three Dimensions of Engineering DevOps—People, Process, and Technology 55
Figure 7 DevOps Engineering 3D Game 60

Figure 8 DevOps Engineering Cube Puzzle 60
Figure 9 DevOps Value-Stream Pipeline Example 62
Figure 10 Value-Stream Map Template 63
Figure 11 Value-Stream Map Example 64
Figure 12 Value-Stream Map Lead Time Example 65
Figure 13 Value-Stream Map—Quality Example 65
Figure 14 Value-Stream Map—Future State Example 67
Figure 15 DevOps Value-Stream Simulation 68
Figure 16 Value-Stream Simulation Results 68
Figure 17 Value-Stream Management Blueprint 72
Figure 18 Value-Stream Management Tool 73
Figure 19 Application Release Automation (ARA) Engineering Blueprint 79
Figure 20 DevOps Version Management Blueprint 85
Figure 21 Continuous Security Engineering Blueprint 91
Figure 22 DevOps Service Catalog Examples 100
Figure 23 Continuous Feedback Service Catalog Metrics 101
Figure 24 Continuous Feedback Monitoring Framework 102
Figure 25 Flexible Standardization 102
Figure 26 Flexible Service Catalog Management Workflow 103
Figure 27 DevOps Governance Engineering Blueprint 110
Figure 28 Site Reliability Engineering (SRE) Blueprint 117
Figure 29 DevOps Disaster Mitigation and Recovery Blueprint 122

PART III: ENGINEERING APPLICATIONS, PIPELINES, AND INFRASTRUCTURES FOR DEVOPS

Figure 30 DevOps CI/CD Pipeline 148
Figure 31 CI/CD Tools 150
Figure 32 CI/CD Pipelines for Three-tier App 154
Figure 33 CI/CD Pipeline for Database 156
Figure 34 CI/CD for Microservices Pipelines 157
Figure 35 CI/CD Pipelines in the Clouds 158

Figure 36 CI/CD Pipeline Maturity Levels 162
Figure 37 Containers 169
Figure 38 Containers Workflow 169
Figure 39 Infrastructure-a-Code (IaC) 171
Figure 40 Infrastructure-as-Code Tools 172
Figure 41 DevOps with Hybrid Cloud 182
Figure 42 DevOps Hybrid Cloud Orchestration Stack 183
Figure 43 DevOps Multi-Cloud 188
Figure 44 Multi-Cloud Services Forecast 190
Figure 45 Cloud Service Providers Comparison 190
Figure 46 Cloud Services Comparison 191
Figure 47 Multi-Cloud Architecture 192
Figure 48 Traditional Cloud-Native Apps[RW67] 193
Figure 49 Elastic Infrastructure Maturity Levels 195
Figure 50 Continuous Test Engineering Blueprint 200
Figure 51 Advanced Continuous Test Squeeze 202
Figure 52 Advanced Continuous Test Engineering Blueprint 204
Figure 53 Continuous Monitoring Engineering Blueprint 207
Figure 54 Continuous Delivery and Deployment Engineering Blueprint 216
Figure 55 Blue-Green Deployments 219
Figure 56 Feature Toggles with A/B Testing 220
Figure 57 Feature Flag Roll-out 221
Figure 58 Canary Testing 222
Figure 59 Microservices Deployment 224

PART IV: DEVOPS SEVEN-STEP TRANSFORMATION ENGINEERING BLUEPRINT

Figure 60 DevOps Seven-Step Transformation Engineering Blueprint 228
Figure 61 DevOps Transformation Goal Scorecard Example 238
Figure 62 DevOps Transformation Practices Topics Scorecard Example 239

Figure 63 DevOps Assessment Engineering Blueprint 242
Figure 64 DevOps Practices Maturity Assessment Tool Example 243
Figure 65 DevOps Engineering Solution Requirements Matrix 245
Figure 66 DevOps Transformation Solution Engineering Blueprint 249
Figure 67 Engineering DevOps Transformation RoadMap Example 251
Figure 68 DevOps Backlog Example 251
Figure 69 DevOps ROI Calculation Example 253
Figure 70 Realize DevOps Transformation Engineering Blueprint 257
Figure 71 DevOps Solution Operations Engineering Blueprint 270
Figure 72 DevOps Expansion Engineering Blueprint 275
Figure 73 DevOps Beyond Continuous Improvement Engineering Blueprint 280
Figure 74 DevOps Continuous Learning Engineering Blueprint 285

Introduction

Do you want to know "How to do DevOps"? Are you "doing DevOps" but not satisfied with the results you are getting? Then you've come to the right place. DevOps is complex. It does not come with a prescription, installation guide, user's manual, or maintenance manual. It doesn't even have a standard definition! How can anyone expect to get good results with DevOps without clear and definitive guidance?

There are plenty of books that describe different aspects of DevOps and customer user stories, but up until now there has not been a book that frames DevOps as an engineering problem with a step-by-step engineering solution and a clear list of recommended engineering practices to guide implementors. My book is unique in that it provides a step-by-step engineering prescription that can be followed by leaders and practitioners to understand, assess, define, implement, operationalize, and evolve DevOps for their organization. It provides a unique collection of engineering practices and solutions for DevOps. By confining the scope of the content of the book to the level of engineering practices, the content is applicable to the widest possible range of implementations.

This book was born out of my desire to help others do DevOps, combined with a burning personal frustration. The frustration comes from hearing leaders and practitioners say, "We think we are doing DevOps, but we are not getting the business results we had expected." There are many documented references, courses, and other sources of information that explain a multitude of different aspects of DevOps, but there is not a comprehensive step-by-step prescriptive guide that details how to engineer a DevOps solution for any organization given any starting point. ***Engineering DevOps*** takes a strategic approach, applies engineering implementation discipline, and focuses operational expertise to define and accomplish specific goals for each leg of your DevOps journey.

This book guides the reader through a journey from defining an engineering strategy for DevOps to implementing *The Three Ways of DevOps* maturity using engineering practices: *The First Way* (called *"Continuous Flow"*) to *The Second Way* (called *"Continuous Feedback"*) and finally *The Third Way* (called *"Continuous Improvement"*).

This book does not provide detailed hands-on tool- or software code-specific information. There are many other sources that cover details of tools and software code. The level of detail needed for tools or software code is NOT within the scope or the objective of this book. Specific tools or software code are not included because the capabilities of the specific tools or software code are limited to specific ecosystems, and they more likely will become obsolete because the tools and software code systems evolve rapidly. This book is deliberately intended to be a guide that will continue to be relevant over time as your specific DevOps and DevOps more generally evolves.

Structure of this book

This book is organized as an engineering reference guide presented in five parts as follows:

Part I: What Is Engineering DevOps and Why Is It Important? is organized into three chapters. Chapter 1, "What is Engineering DevOps?," explains engineering concepts and terms used in this book that have proven to work with many clients I have encountered during my consulting experiences. My depiction of a ***DevOps Engineering Blueprint*** is presented for as a useful, practical "big-picture" reference for discussing engineering the major parts of DevOps. Chapter 2, "Nine Pillars of DevOps," describes a classification of DevOps practices that I have found to be useful characterization of DevOps engineering practices that can be readily applied when engineering DevOps implementations. Chapter 3, "Why Is Engineering DevOps Important?," explains the benefits of taking an engineering approach to engineering DevOps.

Part II: Engineering People, Processes and Technologies for DevOps provides a comprehensive explanation of recommended engineering practices for the higher levels of the DevOps Engineering Blueprint. DevOps transformations are not normally understood or applied using strict engineering principles. Engineering practices are presented in nine chapters as follows: "How Should DevOps Be Engineered?" includes a discussion of DevOps Engineering Maturity Levels for each of the Three Dimensions (People, Process, and Technology), Twenty-Seven DevOps Engineering Critical Success Factors, and Lean DevOps Value-Stream Pipeline Engineering. I describe *recommended engineering practices* for the top layers of the DevOps Engineering Blueprint in the next chapters, which include "Value-Stream Management (VSM)," "Application Release Automation (ARA)," "Version Management," "Continuous Security (a.k.a. DevSecOps)," "Service Catalog," "Governance," "Site Reliability Engineering (SRE)," "Disaster Mitigation," and "Recovery."

Part III: Engineering Applications, Pipelines, and Infrastructures Engineered for DevOps provides a comprehensive explanation of recommended engineering practices for the lower levels of the DevOps Engineering Blueprint, presented in six chapters as follows: "DevOps Applications Engineering," "CI/CD Pipelines Engineering," "Elastic Infrastructure Engineering," "Continuous Test Engineering," "Continuous Monitoring Engineering," and "Continuous Delivery and Deployment Engineering."

Part IV: DevOps Seven-Step Transformation Engineering Blueprint provides a description and tools for my approach to realize and evolve DevOps. The seven steps are Visioning, Alignment, Assessment, Solution Engineering, Realization, Operationalize, and Expansion. By using these tools, DevOps leaders and practitioners can create, implement, operate, and expand their DevOps across the organization. The chapter goes further to explain how to evolve DevOps from a successful *First Way* DevOps (*Continuous Flow*) towards realizing more advanced *Second Way* (*Continuous Feedback*) and *Third Way* (*Continuous Improvement*) DevOps implementations. This part includes a discussion of **"Beyond Continuous Improvement"**—a look at emerging technologies that are shaping DevOps in the future and how you can prepare your DevOps and yourself for the future. This part includes a discussion of how to set up an effective DevOps engineering training program that supports continuous learning of DevOps engineering skills needed to maintain and enhance DevOps.

Part V: Appendices, Continuous Learning, and References includes materials and sources that I have found most useful for engineering DevOps.

Conventions That Are Used in This Book

To highlight a key concept, I indent and mark the paragraph with the following prefix: !! Key Concept !!

Example:

> **!! Key Concept !! The following is a key concept worth committing to memory.**

To denote a reference source, listed in the References section of this book, I use a superscript.

Example:

The Phoenix Project[RB3] indicates a reference to item B3 listed in the REFERENCE section of this book.

DevOps itself is not static. DevOps is comprised of an evolving body of knowledge. DevOps engineering practices and solutions are constantly evolving. It is expected that changes to DevOps engineering practices and solutions will evolve faster than new issues of this book can be published. I invite you to my website www.EngineeringDevOps.com to find a complete and current list of DevOps engineering resources. I update the content on my website to keep current with industry recommended engineering practices. From time to time the latest updates will be consolidated, and a new version of the book will be offered.

PART I

WHAT IS ENGINEERING DEVOPS, AND WHY IS IT IMPORTANT?

"Then there entered into the hall the Holy Grail covered with white samite, but there was none might see it, nor who bare it."

—**Le Morte d'Arthur**, *BOOK XIII CHAPTER VII*

"said Sir Gawaine, we have been served this day of what meats and drinks we thought on; but one thing beguiled us, we might not see the Holy Grail, it was so preciously covered. Wherefore I will make here avow, that to-morn, without longer abiding, I shall labour in the quest of the Sangreal."

—**Le Morte d'Arthur**, *BOOK XIII CHAPTER VII*

PART I: "What is Engineering DevOps and Why is it Important?" is organized into three chapters.

Chapter 1, "What is Engineering DevOps?," explains the engineering concepts and terms used in this book, which have been proven to work with many clients I have consulted for. It is important to have an agreed-upon definition for engineering DevOps, at least within an organization that is engineering a DevOps solution, so that work is coordinated using the same set of assumptions to avoid misunderstandings. There is no industry standard definition of DevOps engineering, and that is problematic. This book defines DevOps engineering using methods from my own experiences that may be applied to any organization and provides definitions of DevOps engineering terms used in this book to help anyone who is not familiar with the terms in the context of DevOps. I also share my *DevOps Engineering Blueprint* as a useful, practical "big-picture" reference for engineering the major parts of DevOps.

Chapter 2, "Nine Pillars of DevOps," describes a classification of DevOps practices that I created as an alternative to the traditional CALMS model. I have found that the Nine Pillars approach is a more useful characterization of DevOps practices and can be more readily used in an engineering approach to DevOps implementations.

Chapter 3, "Why Is Engineering DevOps Important?," explains the benefits of taking an engineering approach to DevOps and the costs of not doing DevOps in accordance with sound engineering practices.

1

What Is Engineering DevOps?

Like the Grail, DevOps is too often regarded as an ethereal thing, earnestly sought after and pursued. To do it means treasures and miraculous benefits; satisfying, harmonious production; and competitive advantages in abundance.

On the other hand, engineering refers to the design and construction of something, with a clear understanding of its requirements under specific operating conditions. In other words, engineering depends on having clear specifications and definitions. The lack of a standard definition for DevOps and lack of a clear prescription of the operating practices for implementing DevOps have many people who are charged with the responsibility of engineering DevOps implementations puzzled. Ask a group of experts, "How do you know when you have achieved DevOps?" and you will get many different answers.[RW1, RW64]

This chapter explains some of the key challenges of defining DevOps in concrete terms that can be used for engineering DevOps solutions, as well as the importance for having a definition, at least within the scope of an organization that is engineering DevOps. For definitions of DevOps terms that are most important for engineering DevOps implementations, see Appendix A.

Engineers traditionally create and use a set of large-scale structural drawings, referred to historically as "blueprints," to guide the major elements of design, construction, and operation of complex systems. Engineering blueprints for DevOps and some of the key dimensions, pillars, and practices that are key to DevOps are provided throughout this book. These blueprints illustrate in one "big picture" the major constructs associated with DevOps, pillars and practices, and the relationships between them. **Figure 1—DevOps Engineering Blueprint** is the top-level blueprint used in this document. Individual portions of this blueprint are explained in this section, and it is referred to throughout this book.

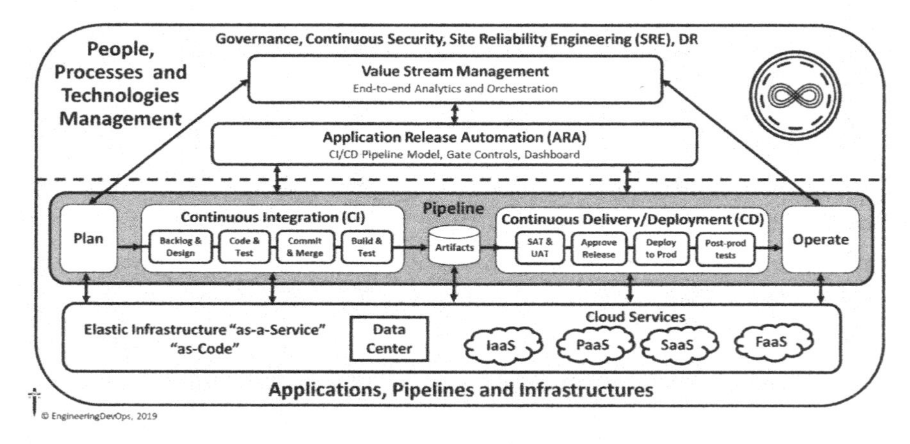

Figure 1—DevOps Engineering Blueprint

DevOps Engineering Blueprint

Engineering requires the entire team to have a common set of instructions. Can you imagine the chaos that would ensue if the stonemasons, ironmongers, and carpenters building Camelot each had their own construction drawings? As they designed and built their components of the structure, they would have become frustrated when their work didn't fit together. King Arthur would not have been pleased either.

There are many ways to label and layout DevOps components and the connections between them. Having a common "big picture" of DevOps components and relationships avoids construction chaos. **Figure 1—DevOps Engineering Blueprint** is the approach I use in my practice and in this book to illustrate the major components associated with DevOps, and the relationships between them, in a manner consistent with recommended engineering practices. Different organizations have different definitions of components and may not have all the components shown in this blueprint, and yet they can still be doing DevOps. What is critical is that the entire team agrees to follow the same DevOps blueprint and strives to work in accordance with shared practices that have been proven to lead to success. All blueprint components are defined in the Appendices and explained in detail later in the book. In this section, for the sake of explaining the blueprint, I will simply refer to the components without going into detail about them.

The background labeled "Governance, Continuous Security, and Site Reliability Engineering" indicates these items apply to all other items in the diagram.

Items positioned higher in the blueprint have higher levels of control and observations than items lower in the blueprint. For example, Value-Stream Management can control and observe Application Release Automation, Pipeline, and Elastic Infrastructure components.

In a real DevOps implementation, the Value-Stream Management component may or may not connect to an Application Release Automation (ARA) component. It may have connections to any of the lower layer components directly. For example, it may connect directly to Elastic Infrastructure components. Similarly, in a real DevOps implementation the ARA component may or may not connect directly to Elastic Infrastructure components.

Items in the component labeled "Pipeline" are shown in a series of "stages" with arrows from left to right. The arrows indicate increasing relative time from stage to stage. The Plan stage is completed before the Continuous Integration (CI) stage, and then the Continuous Delivery/Deployment (CD) stage follows, and finally the Operations stage. The CI and CD stages are further subdivided into stages themselves. The CI stage takes input from the Plan stage and includes stages for Backlog and Design, Code and Test, Commit and Merge, and Build and Test, resulting in a repository of Artifacts depicted by the barrel symbol for database. The CD stage takes as input Artifacts and includes SAT and

UAT, Approve Release, Deploy to Prod, and Post-Prod, resulting in an output to Operations.

The Elastic Infrastructure component has several components noted within it. Data Center and the Cloud components labeled "IaaS," "PaaS," "SaaS," and "FaaS" are not in a time order.

A PowerPoint version of the blueprint can be found on www.EngineeringDevOps.com if you would like to download it or modify it to the suit terminology and layout of DevOps components for your organization.

!! Key Concept !! DevOps Engineering Blueprint

A DevOps Engineering Blueprint that clearly defines abstract labels and layers of DevOps components and their relationships is an important reference diagram to ensure all team members have a common understanding that guides the implementation.

DevOps Engineering Tenets and CALMS

DevOps engineering tenets are the core underlying principles that guide well-engineered DevOps implementations. **Figure 2—CALMS** Model is a conceptual model that explains DevOps tenets. *CALMS*[RW17] is an acronym that stands for Culture, Automation, Lean, Metrics, and Sharing. The CALMS model is useful to explain qualitative tenets of DevOps. It does not, however, provide enough precision to engineer DevOps solutions, and so it is limited in scope. It is important to understand the limitations of the CALMS model to find a more precise model to guide DevOps work.

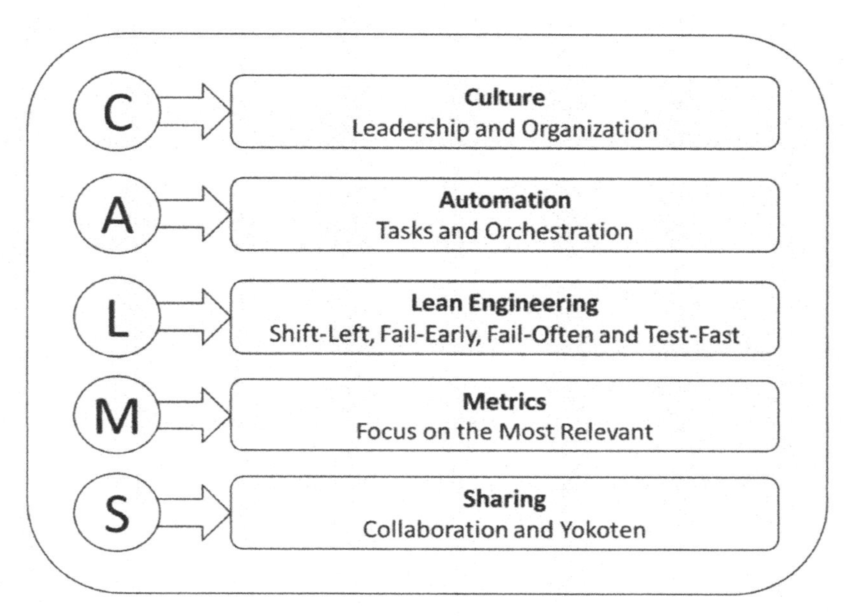

Figure 2—CALMS Model (credited to Jez Humble, co-author of The DevOps Handbook)

Culture—Leadership and Organization: The culture element emphasizes leadership and team organization. DevOps favors cultures that are highly collaborative, leaders that advocate and sponsor DevOps practices, and learning and work arranged with small cross-functional teams with very short communication paths and shared accountability.

Automation and Orchestration: Automation is applied to workflow tasks that can be made faster to remove bottlenecks in the continuous delivery pipeline flow. With DevOps, "Orchestration" refers to a special type of automation that is applied to creating and releasing ephemeral resources. This tenet is key to engineering because all tools in a DevOps toolchain need to support automation capabilities and the ability to integrate with each other in an automated end-to-end workflow.

Lean—Shift-Left, Fail-Early, Fail-Often, and Test-Fast: Lean refers to **lean engineering,**[RW18] which strives to "perfect process "'**flow**'" in a **value**

stream by removing waste (Muda) and minimizing time-to-market (Pull). DevOps tenets "Shift-Left" and "Fail-Early" refer to the lean tenet that doing any work that is necessary to creating value in a value stream is best done as early as possible in the continuous delivery pipeline to avoid that work causing bottlenecks later in the pipeline. Similarly, the tenets of "Fail-Often" and "Fail-Fast" emphasize that test activities are a primary source of waste, or "Muda," for continuous delivery pipelines, so it is best to run tests often with different configurations to detect and repair failures quickly before they become a bottleneck later in the pipeline.

Metrics—Focus on the Most Relevant: Metrics is a critical tenet of engineering DevOps. In DevOps there are many things to measure, including all the elements that make up applications, pipelines, and infrastructures. The challenge is to be precise about what data to measure, how to analyze the data, and what to do with the analysis. Too much information creates noise. Therefore, another key tenet is to engineer metrics, analysis, and communication to ensure that the most relevant information is collected. The most relevant analysis is computed and then communicated to the most relevant people or processes that need to take action as a result of the analysis.

Sharing, Collaboration, and Yokoten: Sharing refers to the tenet that lean engineering recommended engineering practices (*Kaizen*) are proactively shared with other people (*Yokoten*) horizontally across an organization so they can immediately learn from each other without barriers imposed by slower hierarchical organization policies.

Origins of DevOps from an Engineering Point of View

While the word "DevOps" has only been in use since 2009, concepts and practices associated with DevOps have been in existence for many years before.

Contrary to popular myth, the book that most popularized DevOps, *The Phoenix Project*,[RB3] published in 2013, is not the origin of DevOps.

While it may not be an engineering DevOps "how-to" reference manual, it is definitely an entertaining page-turner. The book illustrates how culture, human interactions, and organization silos often undermine the success of IT projects. It also nicely lays out three major phases of DevOps maturity with something called "*The Three Ways,*" which draw from lean manufacturing principles. *The First Way* has to do with realizing "*Continuous Flow.*" This occurs when pipeline stages are connected and operate smoothly without major interruptions. *The Second Way*, "*Continuous Feedback,*" occurs when the *Continuous Flow* pipeline is instrumented with metrics to help identify bottlenecks in the flow. *The Third Way*, which I refer to as "*Continuous Improvement,*" occurs when the pipeline is stable enough to risk some experimentation.

I do agree with many DevOps experts that DevOps completely depends on having the right kind of culture. No technology without the right culture will achieve success with DevOps. I often say that "**culture is a door to Engineering DevOps,**" because if leaders and people don't accept it, then no tool or engineering process will be successful. But without an underlying engineering basis, culture is not sufficient to realize success with DevOps. Nevertheless, I and everyone in the DevOps industry owe a huge debt of gratitude to *The Phoenix Project* and the tireless promotions by its primary author, Gene Kim. Gene is perhaps deserving more credit than anyone for popularizing DevOps through writings, DevOps conference events like the DevOps Enterprise Summit, and motivating other IT leaders to embrace DevOps as a core strategic component at the center of modern digital technology transformations.

A nice article written by Steve Mezak and published on DevOps.com on January 25, 2018, called "The Origins of DevOps: What's in a Name?"[RW2] does a masterful job of explaining how the word "DevOps" came into being during 2009 at the O'Reilly Velocity Conference, during which two Flickr employees (John Allspaw, senior vice president of technical operations, and Paul Hammond, director of engineering) gave a now-famous presentation: "10+ Deploys per Day: Dev and Ops Cooperation at Flickr." There is no need to repeat the whole history here, but I will point out points relevant to this book. The references to lean manufacturing principles and practices is of high importance to DevOps and its links to engineering.

DevOps publications often overlook, or at least gloss over, the extent to which success with DevOps requires, at its core, a strong engineering basis. *The Goal* by Eliyahu M. Goldratt, [RB2] a very entertaining and informative book published in 1984, is a great explanation of lean engineering principles and practices that are so vital to DevOps. *The Goal* predates DevOps and therefore does not correlate specific engineering practices to a DevOps blueprint, but its underlying principles and practices are relevant. This book points out how to correlate them to modern DevOps blueprints.

Going further back, the great "wizard" (statistical quality guru and consultant) W. Edwards Deming is credited with much of the "magic" behind the Japanese industrial recovery including the benchmark lean Toyota Production System. Deming wrote several books, including *Out of the Crisis*, first published in 1982, which is a great summary of scientific and engineering principles and practices that are foundational for lean manufacturing and quality assurance and have been applied to software. *Out of the Crisis* also pre-dates DevOps and does not correlate its engineering principles and practices to the DevOps blueprint as this book does.

My own engineering career, starting in 1974 as an engineering student at Queen's University in Kingston, Ontario, exposed me to software design, building, testing, and deployment—all of which I experienced firsthand. My undergraduate thesis project taught me that testing is an essential part of design and construction. Our small team, consisting of two engineering classmates and I, challenged ourselves to design and implement one of the world's first microprocessor-based private telephone switching systems and some compatible telephone handsets. This we had to do part-time in between other classes and studies. We knew from the start that we had little room for mistakes, as success with the project was essential to graduation! From the start we decided we needed a parallel track to design and implement code and hardware for both "product (switch with handsets)" and "test-ware." This was to make sure we could test the system and verify the operation as we created it. When we finished on time, both the switch and test-ware performed well, and the test-ware served as a great demo tool also. Although the project pre-dated the word DevOps by more than 30 years, it used the same underlying lean engineering practices that DevOps uses. The proof was that we graduated on time!

Glenford Myers' book *The Art of Software Testing*,[RB4] published in 1979, espoused that developers who write software should not test their own code because they would be biased against finding their own failures. In my own experience, I found this idea is not true as indicated by examples in the next paragraph, but nevertheless it took hold in the software industry, at least for functional and system-level quality assurance. Slow, error-prone, back-end loaded "waterfall" processes with large "independent" quality assurance teams became accepted practice. This occurred despite the lessons learned by lean manufacturing that have shown testing should be conducted early to avoid bottlenecks at the back-end of pipelines. It is as if Deming's earlier findings that quality is everybody's responsibility and not to be relegated to some other person was forgotten or ignored "because software is different."

My own experience has shown that **there is an INVERSE relationship between quality of engineering projects and the size of the independent quality assurance team**. I observed while developing and leading projects and quality assurance tools development at Bell-Northern Research (BNR) during the 1980s, and collaborating on other engineering projects since then, that the most efficient projects and highest quality products occur when the engineering team takes responsibility for the end-to-end process, including the quality and deployment stages. To do this requires an end-to-end quality and responsibility mindset. It also requires engineering leadership, training, incentives, and engineering design practices; smart testing methods; and good tools. These added engineering "costs" are too often ignored yet consistently have yielded return on investment (ROI) and advantages for business stakeholders when applied correctly.

Kent Beck's book *Extreme Programming Explained*,[RB5] first published twenty years after The Art of Software Testing in 1999, advocated "test-first development," which finally put testing back in the hands of developers. Test-Driven Development is now a best practice.

Soon after that, *The Manifesto for Agile Software Development* was published in 2001, and "Agile"[RW3] became the software world's favorite new buzzword. Four key tenets of the manifesto are "Individuals and interactions over processes and tools," "Working software over comprehensive documentation," "Customer collaboration over contract negotiation," and "Responding to change over following a plan." I have an opinion that this did little to help the situation from an engineering point of view because many practitioners have interpreted this (incorrectly) to mean, "Software developers should not concern themselves with engineering plans or tests."

The newer *Scaled Agile Framework (SAFE)*,[RW4] initially introduced in 2011 is much improved and espouses development teams to "generate tests for everything—features, stories, and code—ideally before the item is created, or test-first. Test-first applies to both functional requirements (Feature and Stories) as well as non-functional requirements (NFRs) for performance, reliability, etc."

I often say that testing is the primary key to Engineering DevOps because only testing provides sufficient tangible information to measure the quality and completeness of artifacts with engineering precision as they move through the pipeline. Effective lean practices require that testing be "shifted left" as much as possible to the earliest stages of the pipeline to avoid costly bottlenecks caused by latent error detection in later stages.

Jez Humble and David Farley's book *Continuous Delivery*,[RB6] published in 2011, is still regarded as one of the best "how-to" reference books for implementing DevOps, even though it barely mentions the word DevOps and does not use the word "engineering" at all. The book refers to DevOps as a "movement" with the same goal of "encouraging greater collaboration between everyone involved in software delivery to release valuable software faster and more reliably." While the book does meet the goal, it really goes much further by laying out underlying engineering processes for controlling, testing, packaging, releasing, and deploying software changes to production. However, the book does not cover software engineering requirements for planning, design, and post-release "operations," which are also critical for engineering DevOps. Such topics are covered by a scattering of other books and articles. This book emphasizes that a well-engineered, high-performance DevOps requires a complete engineering blueprint with engineering practices that cover the end-to-end set of activities—not just a CI or CD pipeline.

The DevOps Handbook, first published in 2016, is another valuable contribution to the literature that everyone interested in DevOps needs to have. The book includes a wealth of suggestions and case studies for areas beyond Continuous Delivery, including some aspects of software design and, to some extent, operations. However, it does not provide a measurable definition of DevOps, nor does it mention engineering or provide a comprehensive engineering blueprint for implementing DevOps that starts with planning and runs through to operations from any starting point. I have personally worked with many clients who, despite having read all the great published suggestions and case studies, are still confused about what exactly DevOps is and how they should be implementing it in their organizations.

Given this historical perspective, it is my thesis that DevOps can and should be framed as an engineering problem that has an engineering solution. A complete engineering solution requires a blueprint that covers all of the people, process, and technology end to end from conception through to operations, not just a CI/CD pipeline. This is the thesis of this book.

The Dilemma of Defining Engineering DevOps

Business leaders and practitioners need to know, in concrete engineering terms, when DevOps has been accomplished to justify investments that support business goals and how to set limits on projects. Pinning down a universal definition of DevOps that satisfies the diverse community of stakeholders and survives the test of time has proven elusive. DevOps spans a broad and evolving body of knowledge covering multiple human, lean process engineering, and technology disciplines. As explained in *The Phoenix Project*,[RB3] DevOps is never really "done" because the highest level of maturity, "*The Third Way*" (*Continuous Improvement*), defies setting final boundary conditions.

To reconcile this dilemma, some have defined DevOps using vague or transient terms such as a "cultural movement." Others have skirted the problem of definition with long paragraphs describing what is involved to do DevOps, rather than state a definition in concise and concrete terms. Despite these challenges, acceptance of a definition by stakeholders is a key first step in engineering any successful DevOps transformation. Tracking DevOps project progress requires clear definitions that can be measured in concrete engineering terms.

Definition of Engineering DevOps

My definition of Engineering DevOps is broad enough to stand the test of time, yet specific enough to answer key questions addressed by a good definition, including "What is it?," "How it is done?," and "Why do it?" while being concise. When applied to a specific project, the what, how, and why provide tangible things that can be measured in precise engineering terms.

!! Key Concept !! Engineering DevOps Definition

"Engineering DevOps is the application of lean engineering practices (Continuous Flow, Feedback, and Improvement) to the Nine Pillars of Engineering DevOps (Leadership, Collaborative Culture, Design for DevOps, Continuous Integration, Continuous Testing, Elastic Infrastructure, Continuous Monitoring, Continuous Security, and Continuous Delivery) and Three Dimensions (People, Process, and Technology) for the benefit of agility, stability, efficiency, quality, security, availability, and satisfaction."

Whether you create or choose another definition or decide to use mine, it is important that your cross-functional teams agree on one definition on which to align their goals and implementations and set a basis to measure DevOps progress and business outcomes.

DevOps Engineering Terms

There is no official "standard" or comprehensive definition of DevOps terms or categories of DevOps terms. The DevOps Institute, IEEE®, ITIL, ISO, and Wikipedia are good sources for some definitions; however, definitions tend to be different for different sources.

Appendix A of this document provides an alphabetized list of the most popular DevOps terms, abbreviations, and acronyms, with a description consistent with their use in this document.

For a more comprehensive list of DevOps definitions, refer to the "Definitions of DevOps Terms" document posted on my website at www.EngineeringDevOps.com.

2

Nine Pillars of Engineering DevOps

I have great respect for the word ***pillar***. Maybe it is because my dad, Bill, who was the owner-operator of a construction company that included only five people (himself, my mom, my two sisters, and me) often had the occasion to need pillars to prop up roof and other building structures on houses that we were working on. I also respect that my dad was thrifty and resourceful. However, there were occasions that I thought he might be too willing to trade risks to avoid the costs of building tools, such as construction-grade scaffolding. I supposed because I was the only other male in his building enterprise, he sometimes used me as a pillar to hold up building structures while he was laboring to fasten them. I recall, to this day, that my respect for pillars became especially intense on occasions when I was running out of energy as a human pillar, but the thought of having the structure crush me released sufficient adrenaline to counter the effect of lactic acid in my tired teenaged muscles and gave me stamina to be a good pillar. I guess CALMS was not in my lexicon at that time.

That may explain why I like to use the ***pillar*** to describe major "immovable" structural parts of DevOps. It is certainly one reason I decided to study and go into engineering school instead of following my father's footsteps into a life of construction labor. Much like a pillar, a column of a building that helps hold the structure up, a DevOps pillar represents a permanent structural part of any well-engineered DevOps.

In my engineering framework, DevOps is organized in Nine Pillars. These are permanent structural parts of DevOps engineering, no matter where you are trying to do DevOps. As illustrated in **Figure 3—Nine Pillars of DevOps**, the Nine Pillars of DevOps are Leadership, Collaborative Culture, Design for DevOps, Continuous Integration, Continuous Testing, Elastic Infrastructure, Continuous Monitoring, Continuous Security and Continuous Delivery,[RW54, RW55, RW56] each DevOps pillar categorizes specific engineering practices that are useful to describe and evaluate DevOps practices an organization needs to engineer and operate DevOps. CALMS tenets are relevant to each of the pillars. In this section you will find descriptions of each DevOps pillar and examples of associated DevOps engineering practices. I should point out that I have used both seven-pillar and nine-pillar models from time to time. The seven-pillar model incorporates the Leadership pillar into the Collaborative Culture pillar and the Security pillar into the other pillars. I found it better to break out Leadership and Security into separate pillars because they have some distinctive practices that are worth highlighting.

In the diagram you will notice the *Nine Pillars of DevOps* diagram includes two horizontal foundational structures ("Orchestration and Automation" and "Governance") and three roof structures ("*Continuous Delivery CI/CD Pipeline*," "*Application Release Automation*," and "*Value-Stream Management*"). Horizontal foundational and roof structures cross the Nine Pillars because they are relevant to all the vertical pillars. "Orchestration and Automation" is the same as the "A" in *CALMS*. The other horizontal structures are described in later subsections of this chapter.

Leadership Pillar

The DevOps ***Leadership*** Pillar has to do with the aptitudes, attitudes, and actions of people that have leadership roles over teams and organizations that are on a DevOps journey. The following are example practices that have been shown to correlate to well-engineered DevOps Leadership:

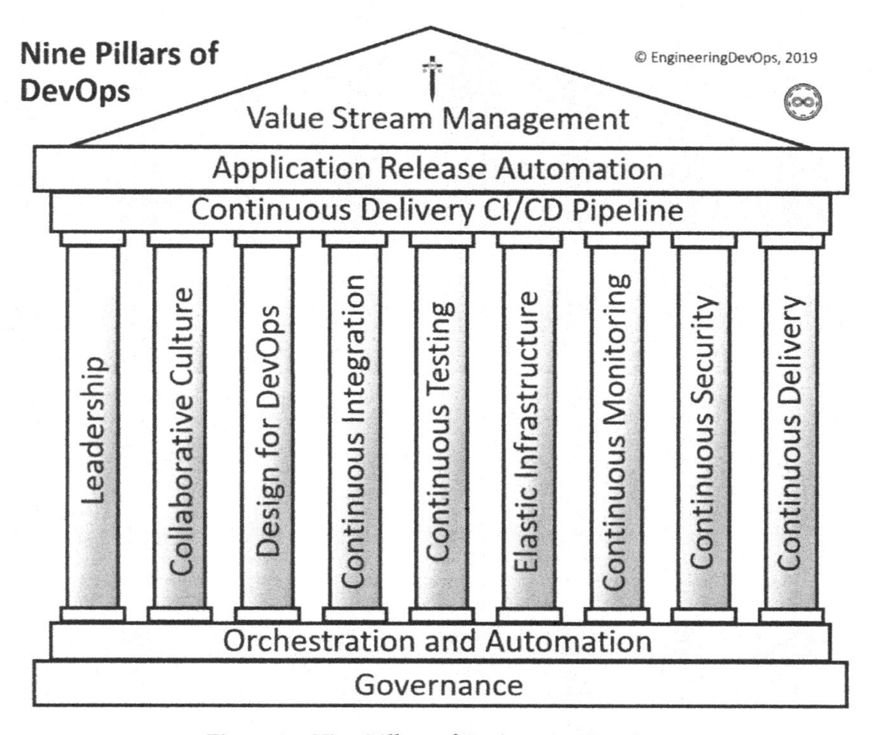

Figure 3—Nine Pillars of Engineering DevOps

- Leadership demonstrates a vision for organizational direction, team direction, and a three-year horizon for team.
- Leaders intellectually stimulate the team and upend the status quo by encouraging and asking new questions and questioning the basic assumptions about the work.
- Leaders provide inspirational communication that inspires pride in being part of the team, says positive things about the team, inspires passion and motivation, and encourages people to see that change brings opportunities.
- Leaders demonstrate supportive style by considering others' personal feelings before acting, being thoughtful of others' personal needs and caring about individuals' interests.

- Leaders promote personal recognition by commending teams for better-than-average work, acknowledging improvements in the quality of work, and personally complimenting individuals' outstanding work.

Collaborative Culture Pillar

The DevOps ***Collaborative Culture*** Pillar has to do with teams and the organization culture within teams. The following are example engineering practices that have been shown to correlate to well-engineered DevOps:

- The culture encourages cross-functional collaboration and shared responsibilities and avoids silos between Dev, Ops and QA, and Product Management.
- The culture encourages learning from failures and cooperation between departments.
- Communication flows fluidly across the end-to-end cross-functional team using collaboration tools where appropriate (e.g., SLACK, HipChat, Yammer, etc.).
- The DevOps system is created by an expert team and reviewed by a coalition of stakeholders including Dev, Ops and QA, and Security.
- Changes to end-to-end DevOps workflows are led by an expert team and reviewed by a coalition of stakeholders including Dev, Ops, and QA.
- DevOps System changes follow a phased process to ensure the changes do not disturb the current DevOps operation. Examples of implementation phases in include: 1) *Proof of Concept* (POC) phase in a test environment; 2) Limited production; and 3) Deployment to all live environments.
- *Key Performance Indicators* (KPIs) are set and monitored by the entire team to validate the performance of the end-to-end DevOps system always. KPIs include the time for a new change to be deployed, the frequency of deliveries, and the number of times changes fail to pass the tests for any stage in the DevOps pipeline.

Design for DevOps Pillar

The ***Design for DevOps*** Pillar has to do with how applications software is designed. The following are example engineering practices that have been shown to correlate to well-engineered DevOps:

- Products are architected to support modular independent packaging, testing, and releases. In other words, the product itself is partitioned into modules with minimal dependencies between modules. In this way, the modules can be built, tested, and released without requiring the entire product to be built, tested, and released all at once.
- Applications are architected as modular, immutable microservices ready for deployment in cloud infrastructures in accordance with the tenets of twelve-factor apps, rather than monolithic, mutable architectures.
- Software source code changes are pre-checked with static analysis tools prior to committing to the integration branch. Static analysis tools are used to ensure the modified source code does not introduce critical software faults such as memory leaks, uninitialized variables, and array-boundary problems.
- Software code changes are pre-checked using peer code reviews prior to committing to the integration/trunk branch.
- Software code changes are pre-checked with dynamic analysis tests prior to committing to the integration/trunk branch to ensure the software performance has not degraded.
- Software changes are integrated in a private environment together with the most recent integration branch version and tested using functional testing prior to committing the software changes to the integration/trunk branch.
- Software features are tagged with software switches (i.e., feature tags or toggles) during check-in to enable selective feature level testing, promotion, and reverts.

- Automated test cases are checked in to the integration branch at the same time code changes are checked in, together with evidence that the tests passed in a pre-flight test environment.
- Developers commit their code changes regularly to trunk—at least once per day.

Continuous Integration (CI) Pillar

The ***Continuous Integration*** Pillar has to do with how changes to software code and build artifacts are "built" or compiled, assembled, or otherwise packaged into executable artifacts for application releases. The following are example engineering practices that have been shown to correlate to well-engineered DevOps:

- A Software Version Management (SVM) system is used to manage all source code versions (e.g., Git, GitHub, Gitlab, Bitbucket, Perforce, Mercurial, Subversion, etc.).
- An Artifact Repository system is used to manage all versions of code images changes used by the build process (e.g., JFrog Artifactory, Nexus, Helix, Archiva, etc.).
- An SVM system is used to manage all versions of pipeline tools and infrastructure-as-code configurations and tests that are used in the build process (e.g., Git, GitHub, Gitlab, Bitbucket, Perforce, Mercurial, Subversion, etc.).
- All production software changes are maintained in a single trunk or integration branch of the code.
- The software versions for supporting each customer release are maintained in a separate release branch to support software updated specific to the release.
- Every software commit automatically triggers a build process for all components of the module that has code changed by the commit.
- Once triggered, the software build process is fully automated and produces build artifacts provided the build time validations are successful.

- The automated build process checks include unit tests.
- Resources for builds are available on demand and never block a build.
- CI builds are fast enough to complete incremental builds in less than an hour.
- The build process and resources for builds scale up and down automatically according the complexity of the change. If a full build is required, the CI system automatically scales horizontally to ensure the build are completed as quickly as possible.

Continuous Testing (CT) Pillar

The ***Continuous Test*** Pillar (my personal favorite) has to do with how tests are used to assess software code and build artifacts changes to ensure they meet the requirements for release. The following are example engineering practices that have been shown to correlate to well-engineered DevOps:

- Development changes are "*Pre-Flight*" tested in a clone of the production environment prior to being integrated to the trunk branch. (Note: "production environment" means "variations of customer configurations of a product.")
- New unit and functional regression tests that are necessary to test a software change are created together with the code and integrated into the trunk branch at the same time the code is. The new tests are then used to test the code after integration.
- "*Green/Blue Deployment*" methods are used to verify deployments in a staging environment before activating the environment to live.
- Release regression tests are automated. At least 85% of the tests are fully automated, and the remaining are auto-assisted if portions must be performed manually.
- Release performance tests are automated to verify that no unacceptable degradations are released.
- "*Canary testing*" methods are used to trial new code versions on selected live environments.

- The entire testing lifecycle, which may include *Pre-Flight*, *Integration*, *Regression*, *Performance*, and *Release Acceptance* tests are automatically orchestrated across the DevOps pipeline. The test suites for each phase include a pre-defined set of tests that may be selected automatically according predefined criteria.
- Test resources are scaled automatically according to the resource requirements of specific tests selected and the available time for testing.
- "*A/B testing*" methods are used together with *Feature Toggles* to trial different versions of code with customers in separate live environments.

Elastic Infrastructure (EI) Pillar

The ***Elastic Infrastructure*** Pillar has to do with how resource requirements (i.e., computing machines, storage and networks") for builds and testing environments vary in near real time depending on the workload requirements to support specific changes and the variable demand of a constantly changing number of users that need to use resources. The following are example engineering practices that have been shown to correlate to well-engineered DevOps:

- The data and executable files needed for building and testing software builds and *Infrastructure-as-Code* are automatically archived frequently and can be reinstated on demand. Archives include all release and integration repositories. If an older version of a build needs to be updated, then the environment for building and testing that version can be retrieved and reinstated on demand and can be accomplished in a short time (minutes to hours).
- Build and test processes are flexible enough to automatically handle a wide variety of exceptions gracefully. If the build or test process for a component is unable to complete, then the process for that failed component is reported and automatically scheduled for analysis, but build and test processes for other components continue.

The reasons for the component failure are automatically analyzed and rescheduled if the reason for the failure can be corrected by the system, but if not, then it is reported and suspended.

- System configuration management and system inventory is stored and maintained in a version-managed *Configuration Management Database* (CMDB).
- Infrastructure changes are managed and automated using configuration management tools that assure *idempotency*.
- Automated tools are used to support immutable infrastructure deployments.
- The user performance experience of the build and test processes by different teams are consistent for all users independent of factors such as location or other factors. Monitoring tools measure user performance experience is consistent for all users in accordance with Service Level Agreements (*SLAs*).
- Fault recovery mechanisms are provided. Build and test system fault monitoring, fault detection, system and data monitoring, and recovery mechanisms that exist. These mechanisms are automated and are consistently verified through simulated failure conditions.
- Infrastructure failure modes are frequently tested using *Chaos engineering*.
- *Disaster Recovery (DR)* procedures are automated and periodically tested.

Continuous Monitoring (CM) Pillar

The ***Continuous Monitoring*** pillar refers to instrumenting, collecting, and analyzing data needed to manage health and performance of applications, databases, pipelines, and infrastructures and to engineer improvements. The following are example practices that have been shown to correlate to well-engineered DevOps:

- Deployment metrics and release gating thresholds continuously monitor all software changes. Example deployment metrics include the following: test completion rate 95%; test pass rate of 99%; any open critical severity bugs mitigated; *MTBF S-curve* is converging and shows consistent improvement over the last three test runs.

- Logging and pro-active alert systems make it easy to detect and correct DevOps system failures. Logs and proactive system alerts are in place for most DevOps component failures and are organized in a manner to quickly identify the highest-priority problems.
- Snapshot and trend results of each metric from each DevOps pipeline stage (e.g., builds, artifacts, tests) are automatically calculated in process and visible to everyone in the Development, QA, and Ops Teams.
- *Key Performance Indicators (KPIs)* for the DevOps infrastructure components are automatically gathered, calculated, and made visible to anyone on the team that subscribes to them. Example metrics are availability (up time) of computing resources for CI, CT, CD processes, time to complete builds, time to complete tests, number of commits that fail, and number of changes that need to be reverted due to serious failures.
- Metrics and thresholds for the DevOps infrastructure components are automatically gathered, calculated, and made visible to anyone on the team that subscribes to them. Example metrics are availability (up-time) of computing resources for CI, CT, CD processes, time to complete builds, time to complete tests, number of commits that fail, and number of changes that need to be reverted due to serious failures.
- Process Analytics are used to monitor and determine improvements for integration, test, and release processes. Descriptive build and test analytics drive process improvements.
- Predictive analytics are used to dynamically adjust DevOps pipeline configurations. For analysis of test results, data may indicate a need to concentrate more testing in areas that have a higher failure trend.

Continuous Security Pillar

The ***Continuous Security*** Pillar, part of what is referred to as DevSecOps, refers to practices for integrating security of applications, databases, pipelines, and infrastructures into the continuous delivery pipeline. The following are example practices that have been shown to correlate to well-engineered DevOps.

- Developers are empowered and trained to take personal responsibility for security.
- Security assurance automation and security monitoring practices are embraced by the organization.
- All information security platforms that are in use expose full functionality via APIs for automatability.
- Proven version control practices and tools are used for all application software, scripts, templates, and blueprints that are used in DevOps environments.
- Immutable infrastructure mindset is adopted to ensure production systems are locked down.
- Security controls are automated so as not to impede DevOps agility.
- Security tools are integrated into the CI/CD pipeline.
- Source code for key intellectual property on build or test machines are only accessible by trusted users with credentials. Build and test scripts do not contain credentials to any system that has intellectual property. Intellectual property is divided such that not all of it exists on the same archive, and each archive has different credentials.

Continuous Delivery (CD) Pillar

The *Continuous Delivery* pillar refers to practices for preparing release artifacts for deployment to production. The following are example practices that have been shown to correlate to well-engineered DevOps:

- Delivery and Deployment stages are separate. The Delivery stage precedes the Deployment pipeline.
- All Deliverables that pass the Delivery metrics are packaged and prepared for Deployment using containers.
- Deliverable packages include sufficient configuration and test data to validate each deployment. Configuration Management tools are used to manage configuration information.
- Deliverables from the Delivery pipeline are automatically pushed to the Deployment pipeline once acceptable delivery measures are achieved.
- Deployment decisions are determined according to pre-determined metrics. The entire deployment process may take hours but usually lasts less than a day.
- Deployments to production environments are staged such that failed deployments can be detected early and the impact to customers can be isolated quickly.
- Deployments are arranged with automated recovery and self-healing capabilities in case a deployment fails.

In summary, the Nine Pillars of DevOps described in this chapter provide a useful prescriptive framework that categorizes specific engineering practices in a way that makes it very clear to engineer DevOps implementations. In the next chapter I explain why the focus on engineering is important to DevOps.

3

Why is Engineering DevOps Important?

Why should you *engineer* DevOps rather than simply "do a DevOps transformation"? As a leader or practitioner of any enterprise, product or service that involves technology, ignoring DevOps engineering may be a career blunder. You may think I jest. After all, DevOps in and of itself is not a product. It is not important to nature or the survival of mankind. It does not fix world hunger, save a species, heal the sick, or even babysit the kids. Yet DevOps's proven track record for delivering a nearly magical array of benefits rightly compels wise technology leaders and practitioners to pursue it because it has power to greatly affect the value of their work and competitiveness for organizations and projects. **In my experience, many organizations have pushed ahead with a DevOps transformation with the expectation of accomplishing the benefits of DevOps and failed because they did not have a clear, end-to-end DevOps engineering blueprint and they did not have or follow an engineering disciplined approach using engineering practices for each of Nine Pillars of DevOps or Three Dimensions (People, Process, and Technology).**

Development and delivery methodologies that preceded DevOps have, when properly implemented and carefully followed, yielded improvements to time-to-market and efficiencies; added stability, quality, and security of products and services; and improved satisfaction for some business, development, or operations stakeholders, compared to the alternative of not using them. However, there were usually trade-offs. Speed versus quality.

Security versus satisfaction. Stability versus efficiency. The magic of engineering DevOps is that it has the power to provide drastic improvements to all these benefit categories at once for all stakeholders, including you. To realize the "magical" benefits of DevOps requires a practical, disciplined, step-by-step engineering approach that is described in this book.

Engineering DevOps Myths and Realities

There are many popular myths surrounding DevOps that are important from a DevOps engineering point of view. This is understandable given that DevOps definitions and engineering practices have not been "nailed down." In the context of this book, it is important to clarify the truth about myths that are relevant to DevOps so that we can see how an engineering perspective can help. These myths can get in the way of seeing the engineering possibilities.

Myth: "DevOps is a cultural movement." The truth is that implementing DevOps requires a careful and balanced engineering approach that considers a wide array of people, process, and technology aspects. While culture is a critical success factor, DevOps is much more than culture. Furthermore, DevOps is not just a movement. In the larger context of software engineering history, the word "DevOps" is new, and perhaps it will be replaced by some other word someday, but the underlying engineering principles and practices that constitute DevOps apply to almost any engineering project throughout the ages.

Myth: "DevOps requires continuous deployments." The truth is that DevOps enables continuous delivery of release artifacts to a staging environment to be ready to deploy safely, but frequent deployments to production is not always a goal. There are valid customer use cases that do not desire frequent deployments to production. For example, customers may not want to receive a new version of a software platform product frequently because each new version may be disruptive and risky for reasons beyond the scope of the product itself. The engineering benefits of delivering complete releases to staging without deployment to production are still considerable. Each release is produced and validated in a pre-production environment and made ready for deployment upon demand without disrupting customers that do not want to take a new release.

Myth: "DevOps applies to any software product, service, or application." The truth is that DevOps engineering yields high ROI in many circumstances, but some cases do not warrant investment in the people, process, and technology changes needed for DevOps. For example:

- Applications that rarely change do not warrant the cost of DevOps changes.
- Organizations unable to change from organization silos will be too frustrating and counterproductive if they try to implement DevOps.
- The application of DevOps to some Commercial Off-the-Shelf (COTS) products may not yield sufficient ROI because business to business barriers may behave as impenetrable organization silos.
- Applications that have few tools in their toolchains suitable for DevOps may find the transition to DevOps too expensive.

Myth: "DevOps requires Agile." The truth is that DevOps and Agile can be complementary or advisories. The notions of working with small incremental changes is common to both Agile and DevOps. However, organizations that embrace Agile are focused on optimization of the development end of the pipeline and may struggle with DevOps, which emphasizes engineering of end-to-end process optimizations.

Myth: "The scope of DevOps is the same as CI/CD." The truth is that a well-engineered DevOps implementation encompass CI/CD but extend beyond integration and delivery into planning and operations also.

Myth: "DevOps does not apply to platform products such as software code embedded in manufactured products." The truth is that DevOps does apply to embedded code products just as well as web services, business applications, and other types of products. It even applies to development of software and services for DevOps itself! In the case of platform products, typically deployments are planned. They are not continuous, but instead continuous deliveries of candidate releases are delivered to a staging environment to make sure the changes that are implemented for the platform are fully developed and verified as the platform is built instead of waiting for a big-bang release.

How Will I Know When I Have Engineered DevOps?

A key tenet of engineering is that processes, projects, and products are specified and measurable. You can tell when you are "done" because when you achieve measurements within specified tolerances those indicate that you are done. This idea of having clear specification and measurable attributes is vital for DevOps processes, projects, and products also. It is completely reasonable for a business manager to ask, **"Why should I invest a specific amount of resources into engineering DevOps projects until progress is measurable and it will be clear when DevOps has been accomplished?"** This is easy to ask, but without some engineering specifications and measurable definitions, it is impossible to provide an answer. To satisfy this requirement I define minimal, measurable conditions that constitute each of "*The Three Ways of DevOps.*"

I define the following to be the minimal, measurable conditions that constitute "*The First Way of DevOps*"—*Continuous Flow:*

- *Continuous Flow* exists from planning to operations without interruptions.
- A defined pipeline implements a defined value stream.
- Outcomes are deterministic and repeatable.
- Each stage in the value stream has measurable exit and entry criteria and gates.
- The work of each stage in the pipeline is defined and bounded.

- Automation is employed in each pipeline stage and between pipeline stages where needed to prevent serious bottlenecks.
- People and manual work may be employed in the pipeline, but delays caused by human interactions with the pipeline are minimal compared to the total time for a change to transit the end-to-end pipeline.
- Processes to identify and handle quality problems and other interruptions to flow are defined.

I define the following to be the minimal, measurable conditions that constitute "*The 2nd Way of DevOps*"—*Continuous Feedback:*

- *The First Way of DevOps—Continuous Flow*, as described in the last paragraph, is in place and stable.
- Metrics for Service Level Indicators, Objectives, and Agreements (SLI, SLO, and SLA) are in place for the application, the pipeline, and the infrastructure. These SLI, SLO, and SLA metrics are the used to direct application release decisions.
- Metrics analysis tools such as dashboards and algorithms are used to aggregate metrics, log data, and produce trends charts for reactive analysis.

I define the following to be the minimal, measurable conditions that constitute "*The 3rd Way of DevOps*"—*Continuous Improvement:*

- *The Second Way of DevOps—Continuous Feedback*, as described in the last paragraph, is in place and stable.
- Retrospectives are routinely conducted for every release to identify improvements.
- Metrics are proactively analyzed to identify improvements for *Continuous Flow* and *Feedback*.
- The organization is routinely searching for and experimenting with new DevOps solutions and improvements through research and industry outreach. By identifying and tracking new solutions, progress towards *Continuous Improvement* becomes measurable.

Benefits of Well-Engineered DevOps

How can DevOps, which does not even have a standard definition, be attributed any benefits at all, let alone the lofty benefits people like me attribute to it?

Thankfully, you don't have to take my word for it. The benefits of DevOps are very well researched and documented in issues of the State of DevOps Report RR1 thanks to the excellent work of Puppet Labs® and the DevOps Research and Assessment Organization (DORA), led by Nicole Forsgren, Ph.D., and her book *Accelerate*,[RB8] co-authored by two of my personal DevOps heroes, Jez Humble and Gene Kim.

Analysis of data presented in the *State of DevOps Report* indicates there are at least six types of measurable benefits reported by high-performance organizations that are using practices associated with more mature DevOps compared to organizations that are using practices associated to less mature DevOps. **The six benefits of DevOps can be summarized in six categories: Agility, Stability, Efficiency, Quality, Security, and Satisfaction.**[RW61] **This granular level of benefit analysis makes it easier to identify goal priorities and specific changes that can be realized with specific engineering practices.**

Agility

The benefit category of Agility indicates the ability of an organization to move and react quickly to produce changes relevant to products and services. The category of Agility, as represented in the *State of DevOps Report*, includes the following measurable benefits:

- Lead-time measured as the duration from one point in the value stream until code is ready to be deployed to production
- Frequency of producing deployable releases to live production
- The percent of time employees spend on new work, such as producing new features or code, compared to time spent on other types of activities

- The extent that product teams break work into small-batch increments
- The extent that workflow is visible throughout the pipeline

The 2016 *State of DevOps Report* presents comparison data showing that high-performing IT organizations indicate they are deploying 200 times more frequently than low-performing organizations and have 2,555 times shorter lead times.

Stability

The benefit category of Stability indicates the extent to which products and services can maintain an operational state despite disturbances. The *State of DevOps Report* indicates the average cost of an outage is $500,000 per hour and can be much higher. The category of Stability, as represented in the *State of DevOps Report*, includes the following measurable benefits:

- Mean-Time-to-Recover (MTTR) from failure/service outages in production
- The percent of code merges from development break the trunk branch

The 2016 *State of DevOps Report* presents comparison data showing that high-performing IT organizations indicate they are achieving 24 times faster recovery from failures and 3 times lower change failure rates than low-performing organizations.

Efficiency

The benefit category of Efficiency is a measure of the ratio of useful output to total input for the processes involved in producing a release of a product or service. The category of Efficiency, as represented in the *State of DevOps Report*, includes the following measurable benefits:

- The percent of time employees spend on all types of unplanned work, including rework
- The extent that comprehensive metrics are available for capital costs of development and operations
- The extent that comprehensive metrics available for keeping track of the non-capital costs of development and operations
- The extent to which lean product management is practiced using highly visible, easy-to-understand presentation formats that show work to be done

The 2016 *State of DevOps Report* presents comparison data showing that high-performing IT organizations indicate they are spending 22% less time spent on unplanned work and 29% more time on new work than low-performing organizations.

Quality

The benefit category of Quality is a measure of excellence of a product or service. In engineering, quality is a measure of deficiencies and significant variations that cause failures against specific customer or user requirements identified during testing and operational experiences. The category of Quality, as represented in the *State of DevOps Report*, includes the following measurable benefits:

- The frequency of failures that require immediate remediation occurring in live production
- The extent that quality tests and test data are sufficient and readily available when needed
- The extent that the organization regularly seeks customer feedback and incorporates the feedback into design

Security

The benefit category of Security indicates the practice of assuring information and managing risks related to the use, processing, storage, and transmission of information or data and the systems and processes used for those purposes. The category of Security, as represented in the *State of DevOps Report*, includes the following measurable benefits:

- Number of times per year that a serious, business-impacting security event occurs
- Number of times per year that an unauthorized user accesses unauthorized information
- Average percent of time that employees spend remediating security issues

The 2016 *State of DevOps Report* presents comparison data showing that high-performing IT organizations indicate 50% less time spent on remediating security issues than low-performing organizations.

Satisfaction

The benefit category of Satisfaction refers to fulfilling the need of employees to feel good about their working environment. The category of Satisfaction, as represented in the *State of DevOps Report*, includes the following measurable benefits:

- The extent that employees are likely to recommend their team as a great place to work
- The extent that employees are likely to recommend the organization as a great place to work
- The extent that the organization culture is practicing good communication flow, cooperation, and trust
- The extent that leaders promote personal and/or team recognition by commending better-than-average work, acknowledging improvements in the quality of work, and personally compliment individuals' outstanding work

The 2016 *State of DevOps Report* presents comparison data showing that high-performing IT organizations indicate they are 2.2 times more likely to recommend their organization as a great place to work and 1.8 times more likely to recommend their team as a great working environment than low-performing organizations.

So how can DevOps do all this? If you doubt the above research analysis, then just consider that DevOps is to software-based businesses what lean manufacturing engineering practices was to the automobile industry when Japanese automobile manufacturers leap-frogged American automobile sales in America during the 1980s and 1990s. The Toyota Production System, the benchmark example for lean practices, reduces waste, increasing efficiency and reducing costs. The high-quality and cost-competitive products Toyota produces are directly linked to Toyota's ability to reduce waste throughout the production process. To this day, Toyota is the auto industry leader in many categories.

!! Key Concept !! Engineering DevOps Benefits

When lean manufacturing engineering practices are applied with engineering precision and discipline to the Nine Pillars of Engineering DevOps and The Three Dimensions, you do indeed get agility, stability, efficiency, quality, security, and satisfaction.

Costs of Not Engineering DevOps Properly

Despite the compelling benefits of DevOps, I suggest that before you dive headlong into a DevOps transformation, you approach DevOps with the utmost respect for engineering. Doing DevOps incorrectly can be just as perilous as not doing DevOps at all.

Since this book uses the Knights of the Round Table and King Arthur as thematic elements, it seems appropriate that we have a remarkable DevOps story of *Knight Capital*. If you are developing a continuous delivery pipeline, you will find the story an interesting read. With scary sub-subtitles like "Attack of the Killer Code Zombies" and "45 Minutes of Hell," it sounds like a script from a fictional Hollywood thriller. Except it wasn't a fiction. It was a documentary. A real horror that ended a top financial firm and caused unrecoverable financial havoc for investors, other firms, and the entire worldwide stock trading industry. Ripple effects of lessons learned from a single event continue to affect the way the trading software is deployed to this day.

A summary of an article posted April 17, 2014 recaps "Knightmare: A DevOps Cautionary Tale."[RW7] Some of the main points from the article are recounted below because this is such an important example of the costs of not engineering DevOps properly. Grab some popcorn and read on.

In 2012 Knight was the largest trader in US equities and managed an average daily trading volume of more than 3.3 billion trades daily, trading over 21 billion dollars... daily. That's no joke!

Between July 27, 2012, and July 31, 2012, Knight manually deployed new software to a limited number of servers per day—eight servers in all. During the deployment of the new code, however, one of Knight's technicians did not copy the new code to one of the eight computer servers.

At 9:30 a.m. EST on August 1, 2012, the markets opened, and Knight began processing orders from broker-dealers on behalf of their customers. The seven servers that had the correct deployment began processing orders correctly. Orders sent to the eighth server triggered a supposable repurposed flag and unexpectedly "brought back from the dead" old code, which began routing child orders for execution but wasn't tracking the number of shares against the parent order—somewhat like an endless loop. Imagine what would happen if you had a system capable of sending automated, high-speed orders into the market without any tracking to see if enough orders had been executed. Yes, it was that bad!

By 9:32 a.m., many people on Wall Street were wondering why it hadn't stopped. This was an eternity in high-speed trading terms. Why hadn't someone hit the kill switch on whatever system was doing this? As it turns out, there was no kill switch. During the first 45-minutes of trading, Knight's executions constituted more than 50% of the trading volume, driving certain stocks up over 10% of their value. As a result, other stocks decreased in value in response to the erroneous trades. During the 45 minutes of hell that Knight experienced, they attempted several countermeasures to try and stop the erroneous trades. There was no kill switch (and no documented procedures for how to react), so they were left trying to diagnose the issue in a live trading environment where eight million shares were being traded every minute. Since they were unable to determine what was causing the erroneous orders, they reacted by uninstalling the new code from the servers it was deployed to correctly. In other words, they removed the working code and left the broken code. This only amplified the issues, causing additional parent orders to activate the bad code on all servers, not just the one that wasn't deployed to correctly. Eventually they were able to stop the system—after 45 minutes of trading.

Altogether, four million bad transactions were executed against 154 stocks total: more than 397 million shares. Knight Capital Group realized a $460 million loss in 45 minutes. Knight only had $365 million in cash and equivalents. In 45 minutes, Knight went from being the largest trader in U.S. equities and a major market maker in the NYSE and NASDAQ to bankrupt. NASDAQ and SEC fines and payments to investors ensued. Needless to say, some employees were let go, and the employability of some were affected.

Okay, it's time to put down the popcorn bowl or clean the popcorn you spilled while squirming in your seat reading about the Knightmare. The entire failure event could have been prevented had the DevOps implementation taken a more comprehensive engineering approach following an end-to-end DevOps blueprint and the engineering practices of the Nine Pillars of Engineering DevOps.

Here are some lessons learned about not engineering DevOps appropriately:

- Put version management of applications, infrastructure, and pipeline code in a version management system.

- Implement and automate a version roll-back and roll-forward process.
- Put circuit breakers in the code and deployment processes.
- Automate deployment—not just the installation but also the recovery procedures.
- Test deployment processes before they are put in use.
- Instrument application code and deployment processes with monitoring tools.
- Make sure all key stakeholders have visibility to release deployment changes and release deployment process activities.
- Train the staff!

I could go on and tell you other interesting "DevOOOPs" stories and engineering lessons learned, but I hope this one is sufficient to impress upon you that you must engineer DevOps properly or you WILL suffer horrible consequences.

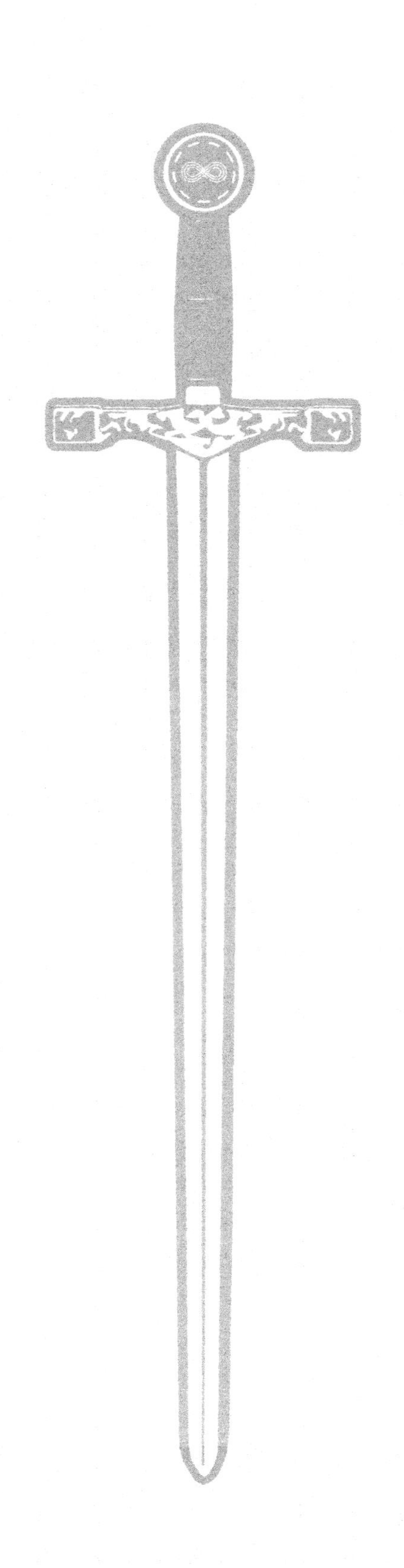

PART II

ENGINEERING PEOPLE, PROCESS, AND TECHNOLOGY FOR DEVOPS

King Arthur rode in the battle exhorting his knights to do well, and himself did as nobly with his hands as was possible a man to do; he drew out Excalibur his sword, and awaited ever whereas the Romans were thickest and most grieved his people, and anon he addressed him on that part, and hew and slew down right, and rescued his people; and he slew a great giant named Galapas, which was a man of an huge quantity and height, he shorted him and smote off both his legs by the knees, saying, Now art thou better of a size to deal with than thou were, and after smote off his head. There Sir Gawaine fought nobly and slew three admirals in that battle. And so did all the knights of the Round Table.

—**Le Morte d'Arthur**, *BOOK V CHAPTER XII*

In this part, **Engineering People, Process, and Technology for DevOps**, a comprehensive explanation of recommended engineering practices for the higher levels of the DevOps Engineering Blueprint is presented in the nine chapters as follows: "How Should DevOps Be Engineered?," "Value-Stream Management (VSM)," "Application Release Automation (ARA)," "Version Management," "Continuous Security (a.k.a. DevSecOps)," "Service Catalog," "Governance," "Site Reliability Engineering (SRE)," and "Disaster Mitigation and Recovery."

4

How Should People, Process, and Technology be *Engineered* for DevOps?

DevOps at its best is engineering greatness. In my experience, well-engineered DevOps always starts by defining a great goal. Greatness is won by careful inspirational leaders, measured tactics, superior technology, a culture of collaboration, courage, and unwavering persistence to applying skills to the task until the goal is attained.

Does DevOps Engineering Require People to be Engineers?

Before we get too far into this chapter, we need to address the elephant in the room. DevOps belongs in the category of ***software engineering*** more than ***computer science***, and that distinction matters to DevOps.

Whether software belongs in the science or engineering category has been the subject of expert debate for many years that continues to this day. Back in 1977, my rural high school guidance counselor was unable to explain the difference between computer science and software engineering sufficiently for me to make a choice of which university program I should apply to. At first, I thought it was simply the lack of knowledge on the part of my high school counselor, so I started my own research efforts to get an answer. After talking to other university counselors and real-world computer scientists and engineers, I determined that there was no consensus.

This is more than just an academic debate. A recent ruling by the Canadian Province of Quebec is a clear example of the way this debate is relevant beyond the halls of academia.

> "Quebec engineers win court battle against Microsoft—The software giant is penalized over its use of the word "engineer" in its professional certification program. Just when you thought it was over, an old debate is reignited."[RW10]

According to the Institute of Electrical and Electronic Engineers (IEEE), software engineering is "the **application** of a systematic, disciplined, quantifiable approach to the development, operation, and maintenance of software"[RS1, RW7].

According to the Association for Computer Machinery (ACM), "Computer science (CS) is the **study** of computers and algorithmic processes, including their principles, their hardware and software designs, their applications, and their impact on society."[RW8]

So is computer science is about "studying," while engineering is about "applying" software? This seems to be an oversimplification based on my own observations of computer science graduates and software engineering graduates working side by side, performing the same jobs in real industry software projects. Nevertheless, both academia and industry continue to distinguish professional software engineering degree programs from computer science degree programs. These degree programs have overlap, but there are key differences in their curricula.

The **"Software Engineering 2014 Curriculum Guidelines for Undergraduate Degree Programs in Software Engineering"**[RR14] was undertaken by a Joint Task Force on Computing Curricula—IEEE Computer Society and Association for Computing Machinery February 2015.RW11 This joint task force consisted of a distinguished panel of engineering and computer scientist experts, which concluded that **"particular attention has been paid to incorporating engineering practices into software development so as to distinguish software engineering curricula from those appropriate to computer science degree programs. Whereas**

scientists observe and study existing behaviors and then develop models to describe them, engineers [do the following]: 1) Engineers use such models as a starting point for designing and developing technologies that enable new forms of behavior. 2) Engineers proceed by making a series of decisions, carefully evaluating options, and choosing an approach at each decision point that is appropriate for the current task in the current context. Appropriateness can be judged by trade-off analysis, which balances costs against benefits. 3) Engineers measure things, and when appropriate, work quantitatively. They calibrate and validate their measurements, and they use approximations based on experience and empirical data. 4) Engineers emphasize the use of a disciplined process when creating and implementing designs and can operate effectively as part of a team in doing so. 5) Engineers can have multiple roles: research, development, design, production, testing, construction, operations, and management in addition to others such as sales, consulting, and teaching. 6) Engineers use tools to apply processes systematically. Therefore, the choice and use of appropriate tools is a key aspect of engineering. 7) Engineers, via their professional societies, advance by the development and validation of principles, standards, and recommended engineering practices. 8) Engineers reuse designs and design artifacts."

To summarize the key points of the above joint task force, the key distinction between software engineering and computer science primarily has to do with methodologies of software design and development. Software engineering methodologies, while having some overlap with computer science, align more perfectly with DevOps methodologies. For designing and developing, software engineers are more apt to use models as a starting point; evaluate each decision point; measure quantitatively; calibrate and validate their measurements; use a disciplined process; have multiple roles; use tools to apply processes systematically; advance by the development and validation of principles, standards, and recommended engineering practices; and reuse designs and design artifacts.

But wait! Most modern software applications use software from open source.[RW12] If DevOps is software engineering, then open source software development must follow software engineering practices. It turns out that it does. According to an interesting article, "Open Source Software Engineering: An Introduction to Open Source Tools,"[RW13] "software engineers make tools and applications that enable users in many domains to perform their work more effectively and efficiently, yet frequently. Thanks to open source, we not only get the source code for development but also the tools to deliver high-quality products."

!! Key Concept !! DevOps Is More Software Engineering than Computer Science

Engineering methods are key to DevOps success. Arguments that software is more of an art form than something that fits engineering disciplines apply to the creative side of developing software products, while DevOps has to do with disciplined methodologies for designing, development, and production of software. A well-engineered DevOps in no way inhibits creativity. Indeed, a well-engineered DevOps facilitates creative design because it reduces production bottlenecks from the software creator.

It is my argument here that DevOps should be classified under software engineering and performed using engineering disciplines. I want to emphasize that this does not mean that only software engineers with an engineering degree should be doing the work of DevOps. I hope that is obvious, but I fear I better clarify that in case some readers think I am a snobbish engineering curmudgeon.

So why does it matter that DevOps is software engineering? The approach to understanding and implementing DevOps from a software engineering perspective emphasizes DevOps blueprint models and specifications, disciplined measurable processes, calibrated tools, systematic progress tracking, principles and recommended engineering practices, validation, strict governance, and artifact reuse.

DevOps has often been referred to as a journey. Not all journeys are alike. Too often I have seen organizations with DevOps journeys that are following a meandering, ad hoc path with few measurable progress milestones, resulting in backtracking or getting lost and not getting to the destination at all. The engineering approach described in this book provides a more certain way to survey the best path, build a clear roadmap based on measurable milestones, and enable a speedy route for the organization to accomplish its goals.

DevOps People, Process, and Technology Engineering Maturity Levels

In the context of engineering, measurements are a critical component. Knowing where you are and where you should go next in an organized, stepwise fashion is a key tenet of engineering. One of the most confusing things about describing how DevOps is engineered is that there are different levels of DevOps maturity, and those levels of maturity are not defined in a standard way. How can you describe a DevOps implementation in any concrete, measurable engineering terms without a measurement guideline?

I am not attempting to define standards for DevOps in this book; however, organizations can and should define their own versions of maturity definitions and then use those to calibrate their own DevOps implementations and progress towards higher levels of maturity.

The Software Engineering Institute[RW14, RS2] defined a Capability Maturity Model (CMM)[RT1] for software with five levels of maturity as depicted in **Figure 4—Capability Maturity Model.**This model, first published in *IEEE Software*[RR4] in March 1988, is a de facto industry standard model for defining the maturity of processes and can be applied to DevOps nicely.

The CMM contains five levels of software process maturity: Initial, Repeatable, Defined, Managed, and Optimizing.

Initial indicates the lack of a stable environment for developing and maintaining software. Few stable software processes are in place, and performance can only be predicted by individual, rather than organizational, capability.

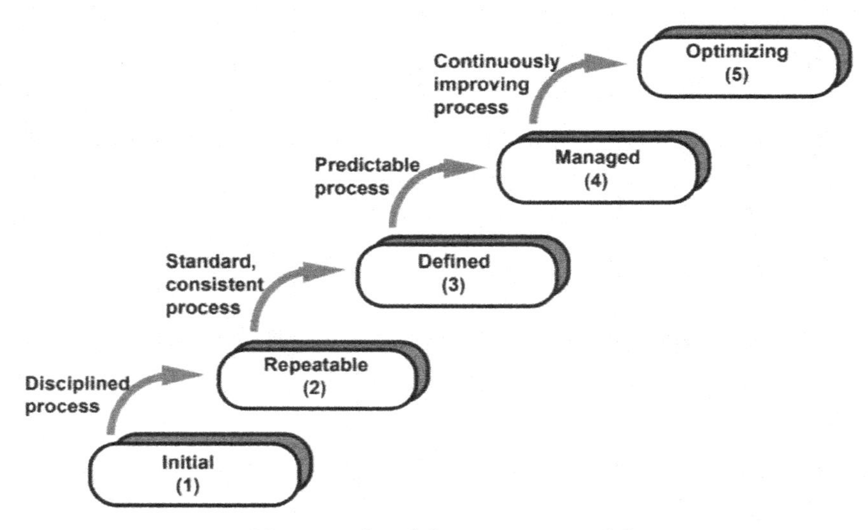

Figure 4—Capability Maturity Model

Repeatable indicates basic level software management controls are in place at the project level. Example measurable elements include the following:

- Software configuration management
- Software quality assurance
- Software subcontract management
- Software project tracking and oversight
- Software project planning
- Requirements management

Defined indicates a standard process for developing and maintaining software is followed across the organization. Example measurable elements include the following:

- Peer reviews
- Intergroup coordination
- Software product engineering
- Integrated software management
- Training program
- Organization process definition
- Organization process focus

Managed indicates quality and processes are measured and analyzed quantitatively across the organization. Example measurable elements include the following:

- Quality management
- Process measurement and analysis

Optimizing indicates the organization is practicing continuous process improvement. Example measurable elements include the following:

- Process change management
- Technology innovation
- Defect prevention

The CMM levels can be used to create a maturity model for DevOps. In doing so, it is important to include measurable elements for the three major dimensions of DevOps—People, Process, and Technology—at each maturity level. Figure 5—Engineering DevOps Maturity Levels is an example of a DevOps maturity model and measurable elements that I use.

Chaos is the initial level of DevOps maturity and very similar to the ***Initial*** level of CMM. There is a lack of a stable environment for developing and maintaining software. Teams are separated into distinct silos with little cross-team communication between them. Few stable software processes are in place, and performance can only be predicted by individual, rather than organizational, capability. Most of the processes including

MATURITY LEVEL	PEOPLE	PROCESS	TECHNOLOGY
Chaos	· Silo team and organization with little communication between silos · Blame and finger-pointing · Dependent on experts	· Requirements, planning, and tracking processes poorly defined and operated manually · Unpredictable and reactive	· Manual builds and deployments · Manual quality assurance · Environment inconsistencies
Continuous Integration	· Managed work backlog · Managed communications between silos · Limited knowledge sharing · Ad hoc training	· Processes defined within silos · No standards for end-to-end processes · Can repeat what is known but can't react to unknowns	· Source code version management · Automated builds, release artifacts, and automated tests · Painful but repeatable releases
Continuous Flow ("First Way" of DevOps)	· DevOps leadership · Collaboration between cross-functional teams · DevOps training program	· End-to-end pipeline automated · Standards across the org for applications, releases processes, and infrastructure	· Toolchain orchestrates and automates builds, tests, and packaging deliverables · Infrastructure is orchestrated as code · Automated metrics and analysis for release acceptance and deployment
Continuous Feedback ("Second Way" of DevOps)	· Collaboration based on shared metrics with a focus on removing bottlenecks · SLIs, SLOs, and SLAs · DevOps Mentors and Guilds	· Proactive monitoring · Metrics collected and analyzed against business goals · Visibility and repeatability	· Applications, pipelines and infrastructure fully instrumented · Metrics and analytics dashboards · Orchestrated deployments with automated rollbacks
Continuous Improvement ("Third Way" of DevOps)	· Culture of continuous experimentation and improvement	· Self-service automation · Risk and cost optimization · High degree of experimentation	· Zero downtime deployments · Immutable infrastructure · Actively enforce resiliency by forcing failures

Figure 5—Engineering DevOps Maturity Levels

software builds and tests and environment setups are manual and error prone.

Continuous Integration is the second level of DevOps maturity, which correlates to the ***Repeatable*** level of CMM. At this level, the focus is to get the front end of the pipeline on solid engineering basis. It would be foolhardy to proceed to automated delivery and deployment until builds and tests and production of release artifacts are repeatable. At this level, there is some cross-team information sharing to support automated builds, release artifacts, and tests sufficient to manage integration.

Continuous Flow is the third level of DevOps maturity, which correlates to the ***Defined*** level of CMM. This level is equivalent to *The First Way of DevOps* discussed in *The Phoenix Project.* At this level, a standard end-to-end process exists in the form of a highly orchestrated and automated pipeline for developing and delivering software. DevOps knowledge and skill levels across the cross-function team are substantial and supported with DevOps training programs. Infrastructure is orchestrated as code. Release acceptance metrics and analysis are automated.

Continuous Feedback is the fourth level of DevOps maturity, which correlates to the ***Managed*** level of CMM. This level is equivalent to *The Second Way of DevOps* discussed in *The Phoenix Project.* People, Process, and Technology dimensions are measured and analyzed quantitatively, and the analytics drive actions such as automated deployments and rollbacks.

Continuous Improvement is the fifth and highest level of DevOps maturity, which corresponds to the **Optimizing** level of CMM. This level is equivalent to *The Third Way of DevOps* discussed in *The Phoenix Project.* At this level, there is a culture of continuous experimentation and improvement. Risks and costs are continuously optimized. Technology solutions support zero downtime deployments, immutable infrastructures, and resiliency.

If these DevOps maturity levels and measurable elements do not fit your organization, that is okay. You are free to create your own if you like. The key point here is that you will need to have a definition of maturity and measurable elements to support an engineering approach to DevOps. And it is critical that you obtain a high level of consensus in your organization before proceeding to use it. Many organizations make the mistake of over-focusing on technologies for DevOps without ensuring associated people and process requirements are kept in step. In this book, I refer to Three Dimensions of Engineering DevOps because all three apply to each of the Nine Pillars of Engineering DevOps.

Three Dimensions of Engineering DevOps—People, Process, and Technology

Besides having a DevOps blueprint, such as the one I presented at the beginning of the book in Figure 1—DevOps Engineering Blueprint, the next step in any DevOps engineering project is to specify "materials" and how they are used to realize a system in accordance with the blueprint. As shown in the **Figure 6—Three Dimensions of Engineering DevOps—People, Process, and Technology**, materials that are required for DevOps come in the categories of People, Process, and Technology. Jennifer Davis and Katherine Daniels in their book *Effective DevOps—Building a Culture of Collaboration, Affinity, and Tooling at Scale*[RB9] state that "successful DevOps culture requires the intersection of people, process, and tools." If you try implement DevOps without considering all three you will be missing key dimensions. Building anything with missing dimensions doesn't work!

People—Leadership, Organization, Teams, and Culture

The People dimension of DevOps includes all the human elements that make DevOps successful.[RW57] This includes leadership, organization, teams, and culture. Effective DevOps requires "an organization that embraces culture change to affect how individuals think about work, value all

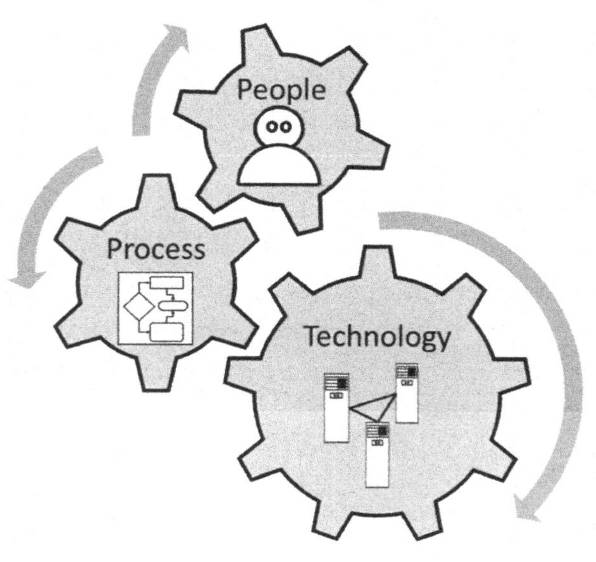

Figure 6—Three Dimensions of Engineering DevOps—People, Process, and Technology

the different roles that individuals have, accelerate business value, and measure the effects of the change."[RB10] Conway's Law[RW16] (named after computer scientist Melvin Conway) indicates that **software tends to reflect the organization structure that created it**. Organizations that have trouble communication between team members will generally create software that has communication problems. The following paragraphs characterize the People dimension of DevOps for each of the five maturity levels:

- At the **Chaos level** of DevOps maturity, teams and organizations exist within separate departments (silos) with little communication between them. There is a little clarity regarding tasks that require shared accountability. This results in a culture of finger-pointing and blame. There is an overdependence on individuals and experts to perform critical tasks. Training and cross-training are not high priorities. Leadership is primarily focused on local department goals and spends little time bridging across departments.
- At the **Continuous Integration level of DevOps** maturity, leadership is proactively involved in directing communications between

department silos, especially for activities that involve cross-functional participation, including peer reviews, build artifact testing, and software integration. Workload backlogs are systematically managed. Cross-team training is a priority but generally not formalized.

- At the ***Continuous Flow* level** of DevOps maturity, *The First Way* of DevOps, leadership visibly and proactively advocates for and sponsors DevOps practices. Collaboration between cross-functional teams is engrained in the culture. DevOps training programs are formalized to ensure team members have DevOps knowledge and skills.
- At the ***Continuous Feedback* level** of DevOps maturity, *The Second Way* of DevOps, the culture is largely self-directed, using shared metrics with a focus on removing bottlenecks. Service Level Indicators (SLIs), Service Level Objectives (SLOs), and Service Level Agreements (SLAs) are used to measure the performance of DevOps systems. The data results drive actions. DevOps training involves the use of advanced concepts such as mentors and guilds.
- At the ***Continuous Improvement* level** of DevOps maturity, *The Third Way* of DevOps, there is a culture of continuous experimentation and improvement. There is a prevailing confidence in the organization and its ability to deliver products quickly and without risk. Training programs that emphasize mastery of the DevOps craft, including reaching out of the organization for expertise, bringing in industry experts, and proactive participation in industry events, is strongly encouraged.

Process—Value Streams and Workflows

The Process dimension of DevOps includes all the workflows that support the pipeline of activities required for the end-to-end value stream of an organization. A **value stream** is the series of activities creating a flow of **value** realized by a product or service that the customer gets. Workflow is a sequence of processes through which a piece of work passes from initiation to completion. The following paragraphs characterize the Process dimension of DevOps for each of the five maturity levels:

- At the **Chaos level** of DevOps maturity, requirements, planning, and tracking processes are poorly defined and operated manually. The results are unpredictable, and systems—where they exist—are designed to be reactive.
- At the **Continuous Integration level** of DevOps maturity, processes that implement workflows are defined within the silos within which they operate to support software development, testing, and integration. There are no standards for end-to-end processes. Systems can repeat what is known, but reaction to unknown circumstances require human intervention.
- At the ***Continuous Flow*** **level** of DevOps maturity, *The First Way* of DevOps, value streams are defined and processes that make up the end-to-end continuous delivery pipeline are automated. Standards exist for cross-functional workflows across the organization for applications, release processes, and infrastructure.
- At the ***Continuous Feedback*** **level** of DevOps maturity, *The Second Way* of DevOps, proactive monitoring systems are in place for all the key elements of the value stream. Metrics are systematically collected and analyzed against business goals. Performance indicators are visible and used proactively to manage processes for applications, pipelines, and infrastructure elements.
- At the ***Continuous Improvement*** **level** of DevOps maturity, *The Third Way* of DevOps, self-service automation is available to developers. Risk and cost optimization are measured and proactively managed. There is a high degree of experimentation with new processes to improve performance.

Technology—Products and Tools

The Technology dimension of DevOps includes technical design and capabilities for products and services; infrastructures; and the tools that are used to plan, create, build, test, package, deploy, and support them. The following paragraphs characterize the Technology dimension of DevOps for each of the five maturity levels:

- At the **Chaos level** of DevOps, maturity tools tend to be "hand tools" rather than production-grade automated tools. Infrastructures and tools that are used to support planning, designs, builds, tests, and deployments are typically are operated manually, with few (if any) disciplined playbooks. The lack of automation or concrete playbooks results in variances and inconsistent results.
- At the **Continuous Integration level** of DevOps maturity, application source code is maintained with a version management system. Software builds, the production of images and release artifacts, and tests are largely automated. This results in repeatable releases, but the release and deployment end of the pipeline requires manual effort, which is error-prone and painful.
- At the ***Continuous Flow*** **level** of DevOps maturity, *The First Way* of DevOps, a CI/CD toolchain orchestrates and automates builds and tests and packaging deliverables. Infrastructure is orchestrated as code. Products, services, and tools are instrumented to report metrics as logs. Automated metrics and analysis are used extensively for release acceptance and deployment.
- At the ***Continuous Feedback*** **level** of DevOps maturity, *The Second Way* of DevOps, applications, pipelines, and infrastructure components are fully instrumented. Activities are driven by metrics and analytics dashboards. Deployments and roll-backs are orchestrated and automated using metrics and analytics to guide deployment and roll-back decisions.
- At the ***Continuous Improvement*** **level** of DevOps maturity, *The Third Way* of DevOps, deployments are using zero downtime methods such as Green/Blue, A/B, and Canary methods. Infrastructures are immutable using containers (such as Docker) and cluster deployment tools (such as Kubernetes). Resiliency of pipelines and infrastructures is actively tested for by forcing failures and enforced by using policies and automation.

Twenty-Seven DevOps Engineering Critical Success Factors

The components to implement the Nine Pillars of DevOps for each of the Three Dimensions of DevOps need to be appropriate for the level of maturity that is being implemented, and at any time they must be kept in a balance. This seems obvious when you think about it. Did you ever see a gold toilet in an outhouse, or an outhouse in a five-star hotel? Well-engineered systems are designed carefully to strike an optimal balance between different factors to accomplish the goals of the project. Engineers of King Arthur's Camelot selected materials and construction methods to suit the requirements for a strong fortification, which were quite different than the materials and workmanship for a squire's house in the village. The dimensions and pillars were large and strong enough to suit the purpose, but not so large that the cost of construction would have been excessive.

It amazes me how many organizations busy themselves over-engineering some parts of DevOps at the expense of others—and then they wonder why the end-to-end system is not balanced. I think this occurs when the implementers have a siloed perspective of DevOps rather than a big-picture engineering blueprint in mind and a clear understanding of the pillars, dimensions, and maturity levels of DevOps needed to clearly define a well-engineered solution.

As an aide-memoire (I took French in school—are you not impressed? Oui ou non?), think of engineering DevOps like a three-dimensional game, as shown in the **Figure 7—DevOps Engineering 3D Game**. The game levels are the Three Dimensions (People, Process, and Technology). You get twenty-seven pieces to place on squares of the game. The game is "won" when all the pieces for all the pillars, for each of the dimensions, line up within the target maturity level. It is also okay if you have some pieces at higher levels of maturity, but that's not desired because that's over-engineering.

Another way to think about the correct relationship between the Nine Pillars and Three Dimensions of well-engineered DevOps for any one maturity level is the cube puzzle illustrated in the **Figure 8—DevOps Engineering Cube Puzzle.**The puzzle is "solved" when each of the Nine

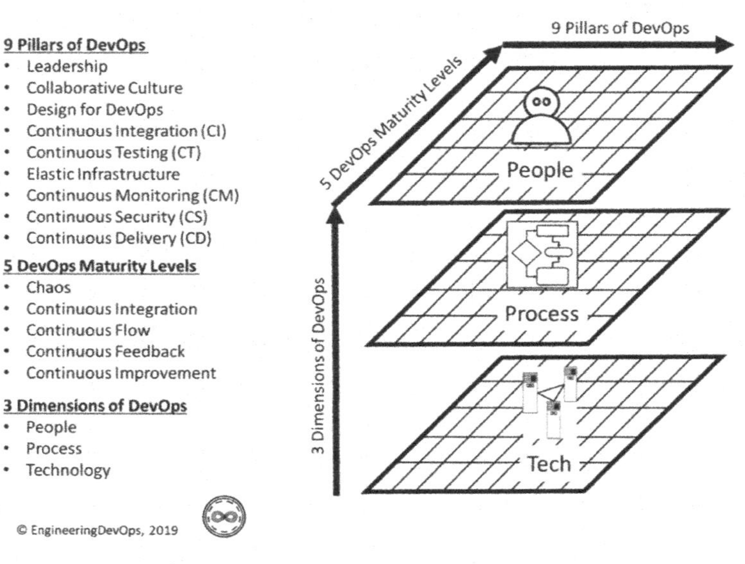

Figure 7—DevOps Engineering 3D Game

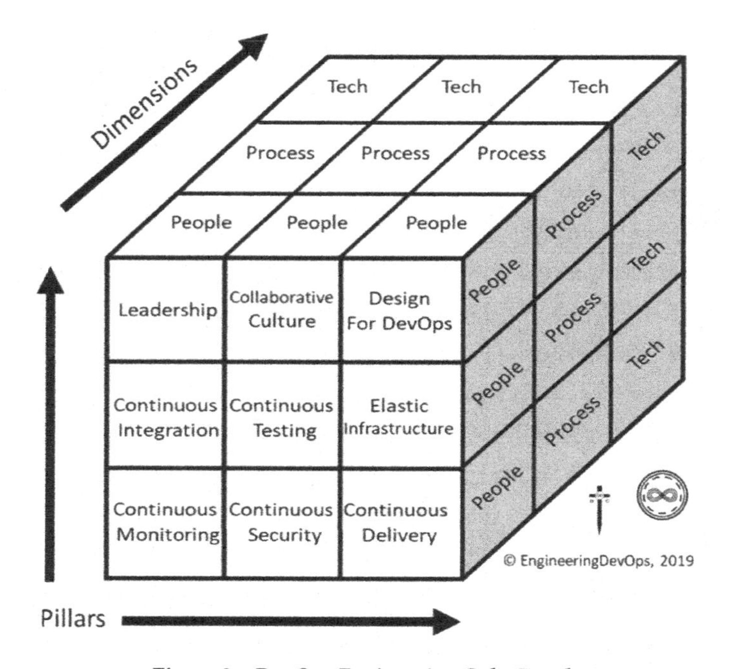

Figure 8—DevOps Engineering Cube Puzzle

Pillars are in line with the Three Dimensions. If you have specifications for DevOps pillars that are missing any DevOps dimension, then you do not have a solution! The correct solution resolves to twenty-seven combinations of Nine Pillars times Three Dimensions. I refer to these as twenty-seven DevOps engineering critical success factors, because the omission of even a single factor jeopardizes the solution.

I know what some (most?) of you are thinking. "Twenty-seven things! I'm not good at solving cube puzzles. I have trouble keeping track of any more than three things. Coffee, milk, and sugar is enough to get right each morning. How can I be expected to balance twenty-seven things?!" My answer is simple, folks. You want the benefits? You deal with it. Do you really think leaders and practitioners of complex engineering projects like designing and building castles, airplanes, boats, web services, and almost every other technology-centric professional doesn't have to engineer at least twenty-seven things to get their products and services right? Get on with it! With this book, you have the DevOps instruction manual in your hand. Read on to continue your noble quest to become master of the game of engineering your lean DevOps machine. In the next section, we will look at a lean value stream pipeline engineering techniques approach to breaking down complex projects into executable project components.

Lean DevOps Value-Stream Pipeline Engineering

Modern supermarkets and shopping malls are amazing places. The giant array of products for affordable prices, all in one place, makes the average consumer take for granted their abundance and convenience. But how did all these products get there? How did they come into existence at all?

In the movie *Six Days and Seven Nights,*[RW19] co-stars Anne Heche and Harrison Ford crash land on a deserted island. While lamenting the lack of modern conveniences, Anne says to Harrison, "Aren't you one of those guys? You know those guys with skills. Yeah, you send them out into the wilderness with a pocket knife and a Q-tip and they build you a shopping mall. You can't do that?" I love that quote. It makes you stop and think about cool stuff like invention, design, supply chains, production, distribution networks, and the incredible number of engineering steps involved to move a product from idea to the cash register—a long chain in which each step adds value. The DevOps continuous delivery pipeline implements a value stream for software. **Figure 9—DevOps Value-Stream Pipeline Example** illustrates this idea.

Not all value streams and pipelines are the same. There is no standard. A tool for visualizing value streams in an easy-to-understand series of process steps has been used in lean manufacturing and business process engineering for years. It is called Value-Stream Mapping.[RB11] Value-Stream Mapping helps identify things important to understanding how DevOps works.

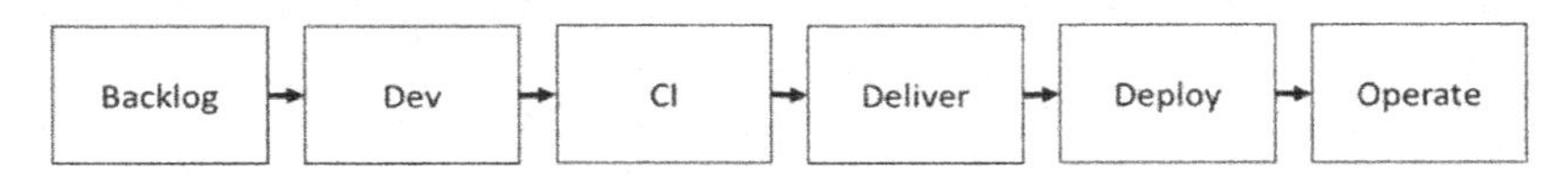

Figure 9—DevOps Value-Stream Pipeline Example

Wouldn't it be great if you could just think about a product and voila, the perfect product appears on the store shelf in front of you? You're a dreamer, too! Unfortunately, in the real world it takes time. **Figure 10—Value-Stream Map Template** illustrates some of the key ideas for understanding how value-stream maps are used to identify bottlenecks in the value stream for DevOps pipelines. Bottlenecks come in various flavours, and none of them taste good. ***Lead time*** and ***quality*** of process steps and series of process steps are two of the most interesting bottlenecks that value stream engineering is concerned with.

Lead Time (LT) is the duration that a process requires from entry to exit. *Process Time (PT)* is the time it takes to perform the work of the process. PT is usually less than LT because processes are rarely 100% efficient. Within a process, there are often things happening that are considered "wasted time" from a process point of view. An example could be a lunch break. In DevOps, waste could be internal wait times for resources to be orchestrated before a build or test can be performed. There can also be waste between process steps. This is referred to *Non-Value-Time (NVT)* in the diagram. Examples of NVT for DevOps value stream is the time waiting for a meeting to get approvals to allow software changes to be promoted to the next stage in the pipeline. Did we forget about quality already? Shame on you. Fortunately, values stream maps never forget. *%C/A* is a ratio, expressed as a percentage. It is used to measure processes that use the results of the prior process without requiring rework. What do C and A stand for? I don't remember, but it's not important. What is important is that rework is a form of waste and a bottleneck that tastes bad to consumers.

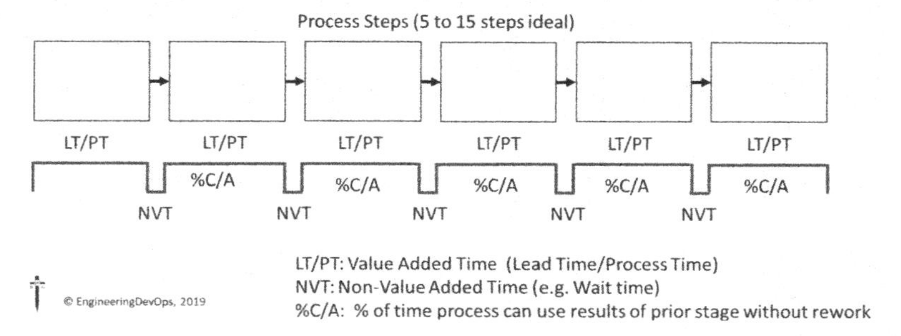

Figure 10—Value-Stream Map Template

Let's look at a real example of how this works. **Figure 11—Value-Stream Map Example** is an actual value-stream map that I worked on together with a team that is responsible for a policy management software application of a large insurance company. Each stage of the value-stream map identifies the process stages, the content of each stage, and the *LT*, *PT*, *NVT*, and *%C/A* values. It sounds easy when I say that in one sentence like that. Getting to this simple map takes a lot of work! A glance at the value-stream map shows where there are key bottlenecks. Or does it? The length of time for each process is only part of the story. Waste is something else. Let's think about this.

A simple calculation *(LT-PT)/LT* for each stage is the percent efficiency for each stage. By adding all the *LT* and *PT* and *NVT* values, you can calculate end-to-end percent efficiency. The numbers highlighted in **Figure 12—Value-Stream Map Lead Time Example** are significant to lead time improvement because they point out processes and interprocess gaps that have a high level of waste and a low level of inefficiency. These are areas to target solutions to reduce waste and improve overall lead time.

The numbers and processes highlighted in the **Figure 13—Value-Stream Map—Quality Example** are important to quality improvement. Remember that *%C/A*, whatever it stands for, is a measure of the percentage of time that a process stage can use results of prior stage without rework. The series of *%C/A* values can be multiplied to get an end-to-end *%C/A* value. In this example, only 15% of the changes committed at the entry gate to the Dev process stage make it all the way to the Operate process stage without requiring rework. What a waste! What can be done? Looking at which process stages are causing the most rework is a good place to look for solutions.

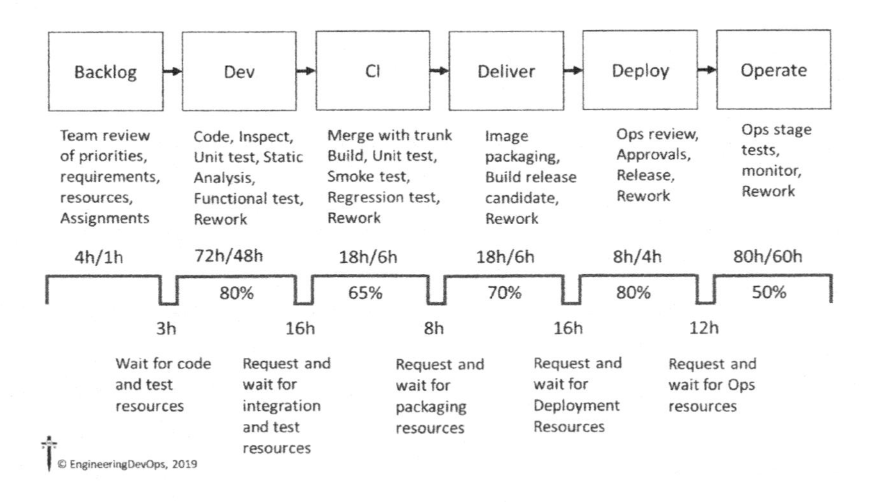

Figure 11—Value-Stream Map Example

So how is this used for DevOps? Let's say you have a specific DevOps goal to reduce your overall lead time from 255 hours to 200 hours per release (55 hours less than the current state) while improving quality by increasing the percentage of changes that do not require rework from 15% to 25% (10 points of improvement from current state). The Value-Stream Analysis tells you which stages and stage transitions to target for process and quality bottlenecks that are going to yield the largest improvements. The current state value stream analysis shown in **Figure 12—Value-Stream Map Lead Time Example** and **Figure 13—Value-Stream Map—Quality**

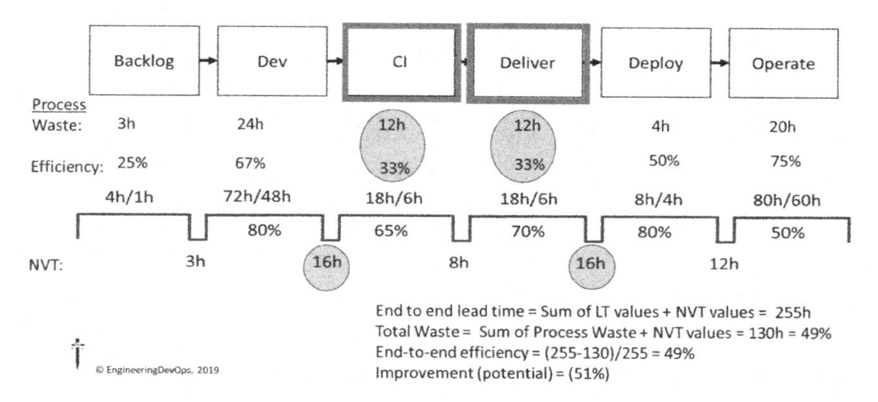

Figure 12—Value-Stream Map Lead Time Example

Example indicate that improvements to CI and Delivery stages and stage hand-offs will most easily yield lead time improvements, while improvements to the Dev, CI, and Deployment stages will most easily yield quality improvements. The value stream current state analysis does not, by itself, tell you specific improvements to make, but it guides you to the areas of the pipeline to best concentrate engineering solutions that will move towards improvement the quickest.

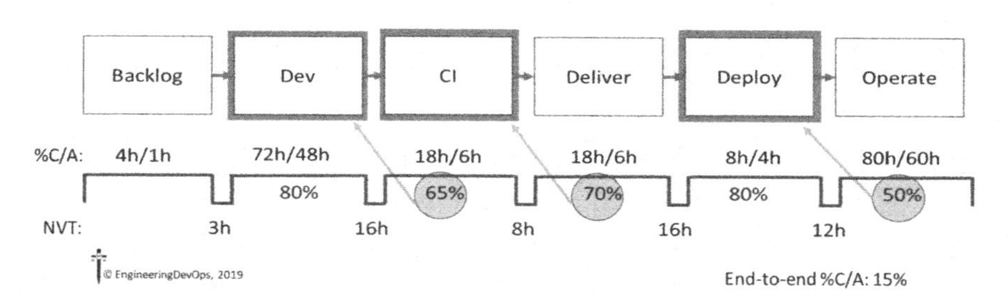

Figure 13—Value-Stream Map—Quality Example

After the current state DevOps value stream is in place, the next step is to create a future state DevOps value-stream map that will describe changes to the pipeline so that it will perform according to the organizations' s DevOps goals. Creating a future state DevOps value-stream map requires DevOps engineering knowledge and expertise to design specific solutions. An example future state value-stream map is shown in the **Figure 14—Value-Stream Map—Future State Example**.

To accomplish the goal of reducing the end-to-end lead time by at least 50 hours, the future state DevOps value-stream map indicates 10 hours lead time reduction to the Dev, CI, and Deliver stages and 6 to 14 hours reduction to most of the stage wait times. To accomplish the goal of improving quality by at least 10 points, the future state DevOps value-stream map indicates %C/A increases of 10 points for CI and Deliver stages and 20 points for the Operate stage, resulting in and end-to-end %C/A of 26% compared to the current state of 15%.

The future state DevOps Value-Stream Map indicates the lead time reduction and quality improvements can be accomplished by implementing the following changes:

- Backlog stage: Include test plan automation tasks in backlog
- Dev stage: Automate unit testing and use Test-Driven Development for functional test creation
- CI stage: Automate CI builds and smoke and regression tests
- Deliver stage: Automate release acceptance tests
- Deploy stage: Automate CAB approvals using ARA
- Between stages: Orchestrate build and test environments; use containers

This list of changes represents a solution from a DevOps expert knowledgeable and experienced in DevOps solutions and recommended engineering practices. There are usually multiple possible solutions to improve DevOps improvements. A DevOps expert considers multiple factors when selecting solutions for a specific value stream and goal. These are discussed in more depth in Part V.

DevOps solutions are implemented incrementally in accordance with DevOps tenets and practices! **Use DevOps for DevOps!** After each increment, results are measured to validate the solution is moving towards the goal. Incremental changes continue until the goal is reached. Once the

DevOps goals are accomplished, new DevOps goals are set and a new current state/future state value stream analysis is conducted.

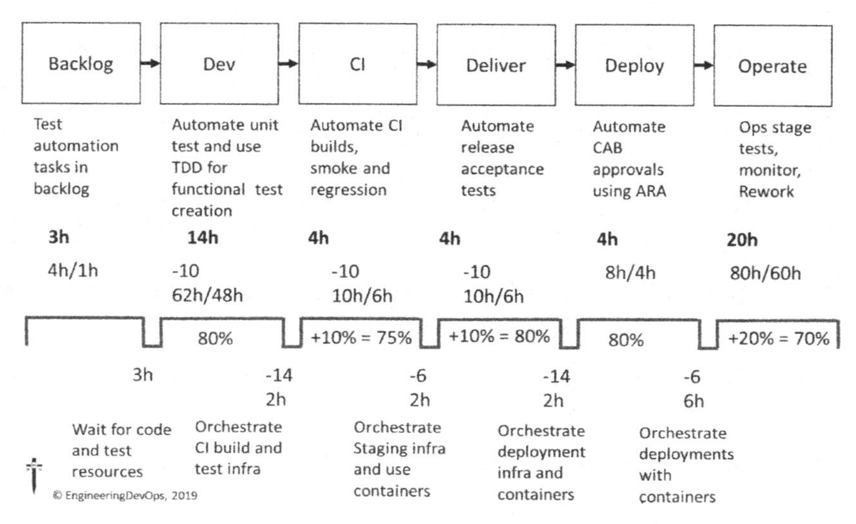

Figure 14—Value-Stream Map—Future State Example

Just for fun, to validate some value stream concepts and some key DevOps tenets, I designed a **Value-Stream Simulator** and a **DevOps machine** that demonstrates these concepts and tenets. To keep this project fun and relatively easy to implement, I designed the simulator and machine to have just enough capability necessary to verify the effects of varying pipeline backlog change rate, stage capacity, stage failure rates on end-to-end lead time, and quality. I did this project using my own spare time in my garage workshop; I finished it on weekends with spare parts on a budget of about $50.

Figure 15—DevOps Value-Stream Simulation illustrates the concept of the Value-Stream Simulator. Each "stage" in the simulator is represented as an abstract process. The backlog state can take new work and rework in the form of tokens at a rate set for each simulation run. The Dev, CI, Deliver, and Deploy stages queue input tokens before processing them at a rate set for each simulation run. Each stage also has a switch to reject a percentage of tokens for rework. The percent rework is set for each stage at the start of each simulation run. The output of each stage consists of all

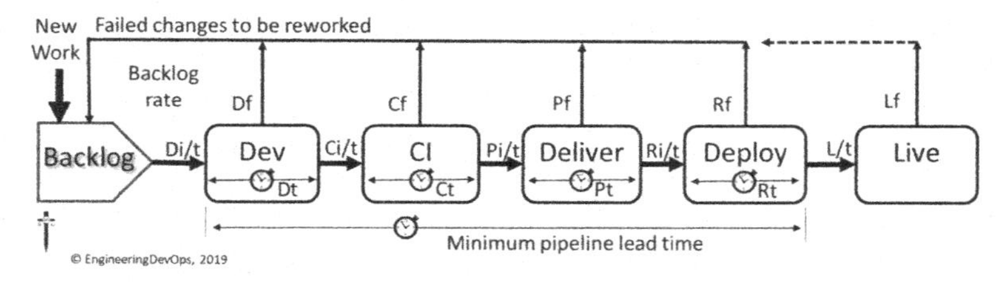

Figure 15—DevOps Value-Stream Simulation

the tokens received minus the ones sent back for rework. The Live stage is like the other stages, except the output is really the end of the simulation and triggers the end-to-end timer clock to stop. The values for total end-to-end lead time, size of release to live stage (measured as the number of tokens passed to live), and total rejected tokens for given settings are used to determine the effects of the changed settings on the pipeline.

The graphs in **Figure 16—Value-Stream Simulation Results** show the results after running the simulator for many settings. It is satisfying to see that more optimum pipeline performance correlated positively for faster pipeline processing (= backlog burn down rates), quality (= defects per stage), and for stages with queue sizes that were similar (= *Continuous Flow*). These results correlate well with DevOps tenets *Fail-Fast, Fail-Early*, and *Continuous Flow* and demonstrate how these tenets affect the performance of the value stream.

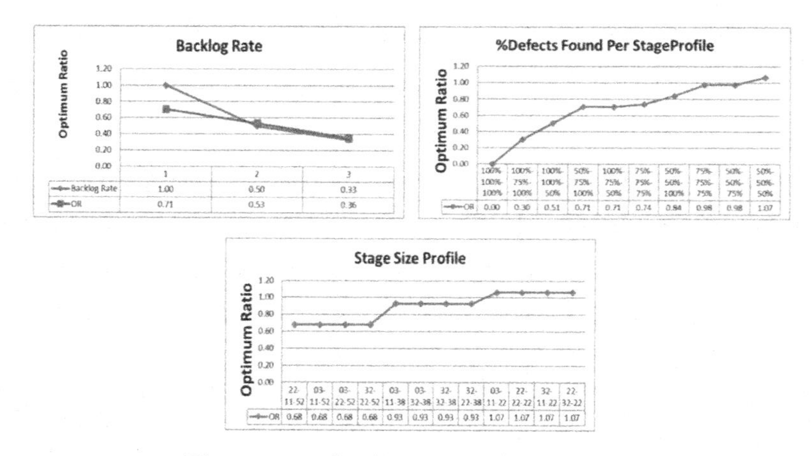

Figure 16—Value-Stream Simulation Results

I created the simulator using Excel. I have on my bucket list the desire and intention to create a more complete simulator using a functional language such as Erlang, Go, or Clojure—something that can simulate multiple parallel interacting microservice pipelines. If you are interested in collaborating on this fun project, please contact me!

If you think the simulator is fun, you should see my **DevOps Machine**. Rube Goldberg would be proud! I won't even attempt to describe it here because it would spoil the fun. If you want a giggle you can see it on SlideShare here: https://www.slideshare.net/MarcHornbeek/devops-machine and YouTube here: https://www.youtube.com/watch?v=SIS5-yVnw_w. I hope you find my DevOps Machine as much fun as I had making it. I am curious to know how you would design and build your own version of a DevOps Machine. I hereby declare the first international DevOps Machine contest open! If yours is better than mine I will send you a signed copy of this book and an official "**DevOps_the_Gray Esq. DevOps Machine Certificate of Excellence.**"

Lean value-stream pipeline engineering sits at the top of the DevOps Engineering Blueprint because this is the layer that governs the entire end-to-end value stream from conception through operations. In the next chapter I explain engineering practices and technologies that assist in managing the end-to-end value stream.

5

Value-Stream Management (VSM)

VSM solutions implement an abstraction layer over one or more end-to-end value streams from planning through to operation and all the stages in the middle.[RW63] VSM is especially powerful when an organization has multiple value streams or pipelines to manage.

Why Is Value-Stream Management Important to DevOps?

VSM solutions address the following problems that are prevalent in "Second Way" and "Third Way" DevOps implementations:

- Fragmented visibility along each pipeline from planning through operations
- Lack of visibility of dependencies across product portfolios and multiple pipelines
- Work-in-progress and work-in-aggregate tracking
- Optimization across disparate pipelines
- Facilitate methodologies for continuous improvement

How Does Value-Stream Management Work with DevOps?

As shown in **Figure 17—Value Stream Management Blueprint**, VSM manages the application delivery process from inception to delivery across the entire enterprise portfolio of multiple value streams. VSM supports DevOps governance, notification, insights, and analytics, as well as orchestration across teams, in a way that would be difficult to accomplish without VSM tools.

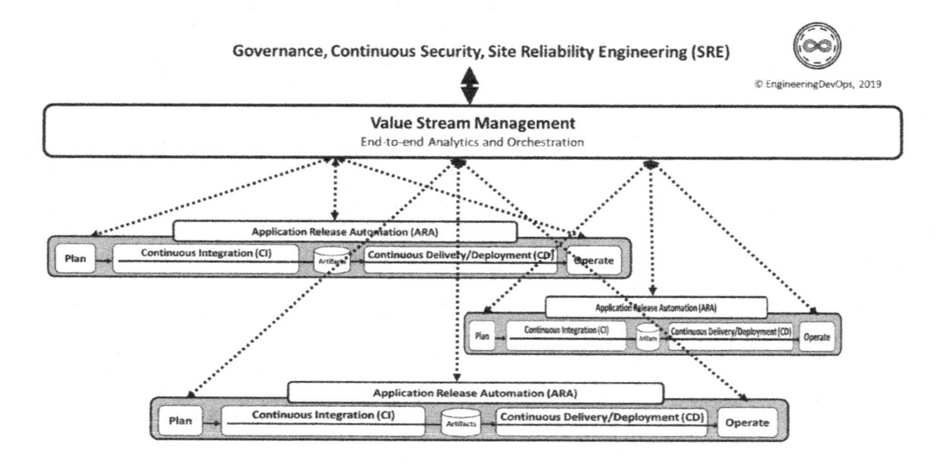

Figure 17—Value Stream Management Blueprint

As shown in **Figure 18—Value-Stream Management Tool**, VSM tools are key to supporting support release planning, coordination and orchestration, and shared environments management taking a broader view of releases to include from planning all the way through deployment and connecting the points in between. It is possible to build these capabilities from ARA or DevOps toolchains, but VSM tools have the advantage of being designed for this right out of the box. VSM tools offer governance and value regardless of the mix of ARA tools or even development methodology, making them an enabling technology for managing the digital transformation journey.

Value-stream management solutions provide the following three key areas of functionality:

1. Integration and Common Data Model. These tools standardize the data from across the entire toolchain and relate data between systems, creating a high-fidelity toolchain relating work between the various groups.
2. Management and Orchestration. Management of key functional areas of the toolchain include functions from enterprise planning, release orchestration, deployment orchestration, environment management, and deployment orchestration. These management functions standardize the workflow regardless of the underlying tooling.
3. Decision-Making and Analytics. VSM tools provide role-based visualizations and analytics into the entirety of the development and delivery process. It enables a focus on delivering value and reducing waste, tracking lead time, process times, and gaps.

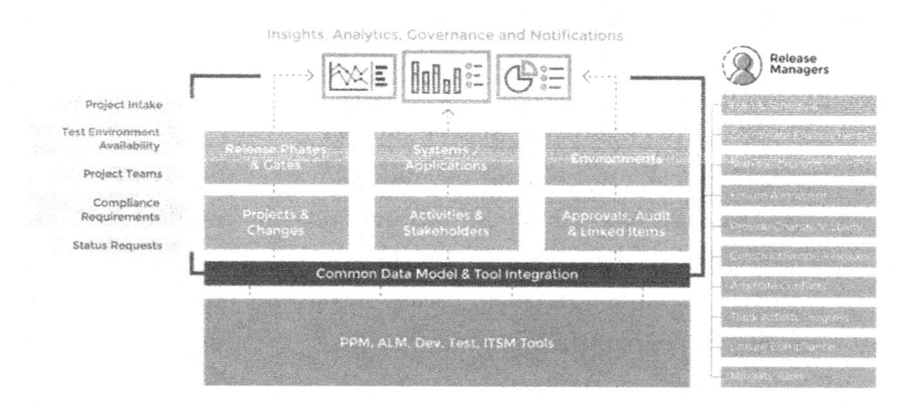

Figure 18—Value-Stream Management Tool

The following are example engineering practices for VSM:

- End-to-end release processes are made visible and controlled using a VSM tool.
- Real time status of releases is made visible using a VSM tool. This enables proactive decision-making around release scope, schedules, and resources to achieve optimal business objectives.

- A real-time dashboard displays the status of all the releases.
- Team members and automated tools can update their release activities in a VSM tool.
- A VSM tool accepts changes to release schedules and provides visibility into the impact of changes to releases.
- Release dependencies are visible using a VSM tool that automatically generates an impact matrix showing releases, application changes, and their scope.
- Compliance audits are supported by a VSM tool that keeps track of activity (who did what, and when) and supports approval workflows.
- Deployment plans, including task assignments and approval workflows, are captured in a VSM.
- Dependencies between deployment steps are captured in a VSM tool.
- A VSM tool tracks deployment task assignment, notifies responsible parties of their tasks, and provides real time status of deployments.
- Execution times of deployment steps are captured in a VSM tool to support analysis for continuous improvement.
- Issues and incidents occurring during deployment are captured in a VSM tool to support analysis for continuous improvement.
- Shared environment conflicts are managed using a VSM tool that provides visibility into environment availability and supports workflows to request/release environments.
- Capacity planning is supported with a VSM tool that captures environment usage and provides utilization reports.
- Environment configurations are captured in a VSM tool.
- Booking requests for environments are handled using a self-service VSM tool.
- Patching and maintenance of environments are supported by a VSM tool that indicates when environments are down.
- Environment change requests are supported by a VSM tool that can pull environment configuration data from existing CMDB, discovery, deploy, or ITSM tools.

- Dependencies between environments are made visible with a VSM tool.
- Compliance audits are supported by a VSM tool that captures environment changes and approvals.

What Is Needed to Engineer a Value-Stream Management Solution for DevOps?

The following steps are recommended to engineer VSM:

1. Conduct a VSM assessment.
   a. Orchestrate leadership interviews.
   b. Leverage discovery tools.
   c. Conduct assessment surveys.
   d. Run value-stream mapping workshops.
   e. Analyze assessment data.
   f. Formulating a strategy and solution roadmap.
   g. Formulate next steps.

2. Select a VSM solution and conduct a POC.
   a. Survey available tools. Gartner Magic Quadrant and Forester are good sources for tool comparison information.
   b. Select a tool or craft a solution with existing tools, design a POC with two or three use cases on one or two model applications.
   c. Deploy the chosen VSM solution with a select number of initial applications on a trial basis. Monitor the trials and make improvements if needed.

3. Deploy the chosen VSM solution(s) to an initial set of production applications.
   a. Complete implementing the remaining high-priority use cases needed for deployment.
   b. Implement training.
   c. Implement success metrics.

4. Operationalize the VSM solution.
    a. Use the solution for a select set of production value streams for at least three releases.
    b. Monitor performance of the VSM solution.
    c. Conduct retrospectives periodically to determine needed improvements.
    d. Continue to implement more use cases as needed for the VMS solution to gain acceptance with more applications.

In this chapter, it was explained how VSM deals with the highest level of abstraction in the DevOps Engineering Blueprint—which includes a portfolio of People, Process, and Technology needed to orchestrate and automate a complete value stream for a product or service. VSM has total end-to-end scope from planning through to operations. The next chapter explains how Application Release Automation (ARA) sits under VSM to orchestrate and automate applications, pipelines, and infrastructures as needed to support release tasks.

6

Application Release Automation (ARA)

Application Release Automation (ARA) solutions realize an abstract layer over *CI/CD pipelines*. ARA solutions have capabilities to model, organize, control, and visualize CI/CD release pipelines.

Why Is Application Release Automation Important?

Well-engineered architectures separate control structures from components that are being controlled because of economies of scale and scope. If every tool in the pipeline required the pipeline architect to orchestrate and automate the pipeline using tool-specific capabilities in the toolchain, then piecing together a pipeline would require a cascade of special cases. Well-engineered ARA solutions provide abstraction for several things that will most likely be different for each class of pipelines and evolve over time, including the following:

- Applications
- Application artifacts
- Configurations data
- Environment artifacts
- Process artifacts

With DevOps there is a goal to continue to improve the pipeline, such as improvements to release velocity, productivity, and continuous delivery frequency. ARA provides an architecture layer that enables this evolution incrementally as needed

Without ARA, the following are example problems that occur with pipelines that do not have well-engineering ARA systems:

- Application quality problems
- Security events
- Pipeline failures
- Interruptive Reverts
- Process delays
- Schedule delays
- Cost overruns
- Audit failures

How Does Application Release Automation Work?

ARA is also often referred to as Application Release Orchestration (ARO), which consists of any number of the following capabilities:

- Deployment automation
- Pipeline management
- Environment management
- Release orchestration

While enablement of these features by each vendor varies, their inclusion in ARO as a toolset confuses the definition of release automation. At its core, ARO provides packaging, versioning, and deployment of applications and their related artifacts. ARO includes broader workflows and can incorporate manual processes. In most cases, VSM vendors also include release orchestration capabilities as part of their offering.

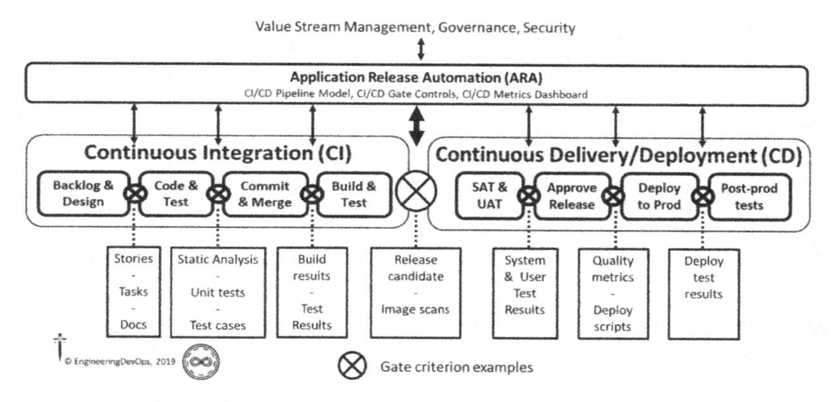

Figure 19—Application Release Automation (ARA) Engineering Blueprint

Figure 19—Application Release Automation (ARA) Engineering Blueprint illustrates the primary capabilities of ARA solutions. The primary components of ARA solutions support the following:

- Pipeline control capabilities in the form of gate criterion as code to control flow between stages
- Environment modeling capabilities, including definition of CI/CD pipeline stages
- Ability to deploy application binaries, packages, or other artifacts to target environments
- Consistent automation for the CI/CD pipeline
- Dashboard and API to make CI/CD metrics visible
- Release coordination with humans, VSM, and governance systems

These capabilities can be realized by piecing together separate tools and augmenting them with scripts. Mature ARA tools have complete ARA capabilities, including the following:

- Automation engine
- Job scheduling features
- Pipeline decision support
- Cloud support
- Ease of use

- Management database
- Agents versus agentless
- High availability
- Integration/plugins

The following are example engineering practices for ARA solutions broken into two broad categories.

ARA capabilities that improve pipeline orchestrate activities are detailed as follows:

- ARA starter templates make onboarding new applications easy.
- The ARA solution supports large proven set of integration plugins popular DevOps tools such as Microsoft tools, Jenkins, GitHub, Docker, JMeter, Go, Python, Java, Artifactory, Jira, ServiceNow, APM, test, build, security, etc.
- The ARA solution supports virtual and cloud infrastructures—Azure, AWS, GCS, OpenShift, Openstack, VMWare, Kubernetes, etc.
- The ARA solution supports configuring pipeline-as-code via declarative languages such as YAML.
- Environments can be defined and referenced abstractly in the pipeline.
- The ARA solution supports blackout periods, maintenance windows, managing conflicts, etc.
- Build and release pipelines are decoupled from applications/artifacts and environments.
- Manual interactions and manual tasks are integrated into pipeline gates.
- Release pipeline templates are reused across teams and release multiple applications.
- Applications are modelled separately from pipelines to simplify managing thousands of artifacts.
- The ARA solution offers a flexible agent architecture to support highly scalable, secure, and configurable configurations.

- Release pipeline and deployment tasks run on premises or cross-cloud with support for network zones, better scalability, security, and simple remote agent "one-click" push install.
- The ARA solution integrates with legacy systems such as mainframe and middleware.

ARA capabilities that make DevOps activities more manageable and visible are detailed as follows:

- Releases and pipelines are not hard-coded to artifacts; target deployments are specified as code.
- Modelling of applications and microservices includes native support for Helm files, Docker Compose, etc.
- The ARA solution user interface is easy to use for application uses.
- The ARA solution integrates with planning features Kanban Board, Backlog, Tasks, Sprints, Jira, Service Now, etc.
- The ARA solution supports capabilities to track usage, inventory, and "what's been released where, when, and by whom."
- Analytics and dashboards are available for Release snapshot status, Release and Deploy Trends, and customized metrics.
- Self-Service Catalog promotes onboarding new users and applications.

What Is Needed to Implement Well-Engineered ARA?

Unless your DevOps environment is very simple, with very few variations of applications, pipeline tools, and infrastructure choices, the key to accomplishing well-engineered ARA solution is architecting a solution around a proven toolset. The primary factors deciding criteria for determining which ARA tools best fit your needs include the following:

- Proven solution
- Ability to support the gate criterion important to your pipelines

- Scale to match the needs of the application and deployment environment
- Licensing model
- Available technical support
- Available training support for users and admin staff
- Integration services
- Sandbox capability to support flexible configurations
- Competitive total cost of ownership
- Single platform, rather than as siloed products
- Easy-to-use, mobile-ready user interface
- DSL to model and execute objects (e.g., application, environment, pipelines, processes, and releases).
- Frequent enhancements
- Supports any scripting language; easy to debug
- Large number of supported DevOps tool plugins (e.g., Jenkins, Docker)
- Agent-based favored for security, scalability, fault tolerance, multiple network zones, and cloud
- SaaS options

One place to look for comparison information is Gartner Magic Quadrant and the Forester Wave. VSM and ARA depend on having versions of artifacts to be managed, orchestrated, and observed. This is only feasible when compatible versions of applications, pipelines, and infrastructures are organized in a version management system. The next chapter explains how this is done in a well-engineered DevOps environment.

7

Version Management

At its bare essence, DevOps is all about version management, because DevOps requires working with incremental versions of software applications, versioned instances of infrastructures, and versioned instances of pipelines. Keeping track of compatible versions of all these increments and instances is essential to be able to piece together a release. Version management is the single source of truth for the product and the company's intellectual property.[RW61] Lose it and you lose the business—and most likely your lunch and future lunch tickets.

I have personally experienced the gut-wrenching pain of being responsible for more than one project that suffered serious setbacks because we were not following recommended engineering practices for version management. In one case we had only one active version management system and one offline backup system. One day, the version management system suffered an unrecoverable storage system failure. It was at that point we discovered the backup system had not been working for the past six months. We lost six months' worth of work. The project was a collaboration with a university. The students were unable to complete their graduate research projects on schedule because recovery took more than four months of manual work.

In another case that I am shy to admit years after the first case (you think I might have learned from the early university project experience), I was running a large engineering department of nearly 400 developers, all of them following the DevOps practice of checking in their code changes to a common version management system daily. A separate IT department was responsible for the servers that hosted the version management systems (one new primary server, one older vintage server used for the replica, and a back-up tape system). As luck would have it, the very week after I bragged to executive management in a quarterly business review that our version management system had worked flawlessly for the past ten years, we suffered a major outage. Thanks, Murphy!

In this case it was far more serious than the prior university project outage, as bad as that was! All the software intellectual property of the company's flagship product was at risk if we could not recover. Furthermore, we had to tell all 400 developers to take an unplanned vacation immediately until we could recover because we knew that it would be a disaster to restart the recovered system with hundreds of parallel software commits. After four days of working around the clock, my engineering team together with IT staff and consultants from two server hardware vendors and the version management software vendor determined the version management system had multiple problems to resolve: the disk array controller on the primary server had and electronics failure, the replication software had a bug so it was replicating incorrectly, and the backup software was not backing up everything needed to complete a restore. What a set of problems!

After receiving and installing a replacement for the failed hardware, we had to run integrity checks on six million lines of source code and many thousands of files of meta data to detect corruptions. We then manually patched the corrupted files, ran system builds and tests to verify a sample of supported product release versions before we could confidently announce the system was restored. Ten days after the system failure, the 400 engineers were back to work, and our DevOps *CI/CD pipeline* was running again. One good thing that came out of this event was that management quickly approved additional replication servers. This was something I had been asking approval for the prior two budget cycles. Thank goodness I have saved copied of those "get out of jail free" emails where management had deferred approvals previously, or my career may not have survived long enough to write about it. Further clean-up was required to acquire, commission, activate, and verify the new replica servers and the updated back-up system and to put in place regular testing of the replica and backup systems. Altogether I calculated the version management failure cost about $2 million unexpected cost for the business unit that year.

Given my two examples it should be clear why version management of source code is essential. With DevOps, version management is much more than keeping source code in a source version management system. As indicated in **Figure 20—DevOps Version Management Blueprint**, versions of your applications, pipelines, and environments need to be managed so that you can diagnose problems with your applications, pipelines, and environment code and data and restore a version of a prior combination if a serious problem requires it. Versions of code and data for applications, pipelines, and environments are typically kept in separate repositories. A master system of record that keeps track of relevant combinations of versions across multiple repositories is necessary. This **multimaster repository** can then be called upon by automated scripts to roll back when needed. Dashboard and real-time alerts ensure the health of the system is visible always. Disaster prevention and recovery structures involve keeping redundant copies of everything, with real-time replication, backups, and procedures automated as much as possible.

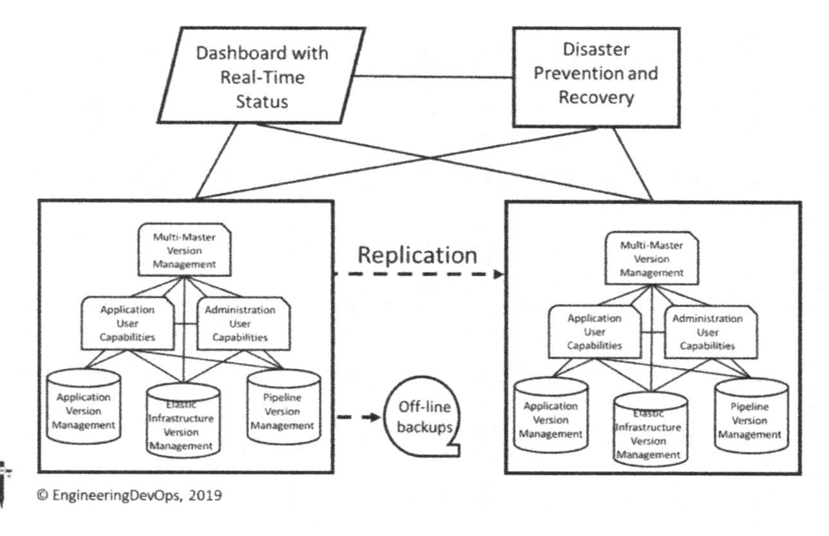

Figure 20—DevOps Version Management Blueprint

The multimaster repository roll-back/roll-forward capability requires version management of metadata and tags for versions of applications, environment/infrastructure, pipelines, documentation, training materials, and tests.

Administration User Capabilities needed include a management user interface and API for controlling and observing the version management system from a system administration point of view, administration of polices for user access groups, KPI metrics, resource quotas and system roll-back/roll-forward versions of applications, environment/infrastructure, and pipeline administration components.

Application User Capabilities needed include user interface and API for controlling and observing version management in accordance with policies for roll-backs and roll-forwards for applications, environment/infrastructure, and pipeline.

Application Version Management repositories need to keep track of source code, build-stacks, images, test-code, circuit-breaker code, documentation, training materials, tests, certificates, metadata, and tags.

Elastic Infrastructure Version Management Capabilities needed include containers registries, build artifacts, configuration management playbooks (e.g., Ansible), network configuration data, storage configuration data, deployment scripts, documentation, training materials, tests, metadata, and tags.

Pipeline Version Management Capabilities needed include release automation scripts, scripts to integrate the pipeline with the elastic infrastructure configurations (e.g., Ansible), build tools, test tools, orchestration interfaces between tools in the toolchain, documentation, training materials, tests, metadata, and tags.

Example version management tools are as follows:

- Git—Git is an open-source tool created by Linus Torvalds, the founder of Linux. Git is a distributed version management system that is very popular with developers because they have direct access to all change history and direct control of their change management workflows, and the response times are much faster than most centralized versions management systems.
- GitHub and GitLab—GitHub and Gitlab are the most popular commercially supported SaaS tools based on Git.
- Perforce—Perforce is a high-performance centralized version management system with capabilities to integrate with Distributed Version Management Systems.
- JFrog Artifactory—JFrog is a release candidate image repository tool that captures versions of release artifacts that are packages needed to create a release.

Example version management engineering practices are as follows:

- Developers check-in software changes to a common trunk branch in version-managed repositories at least once per day.
- All source code and data needed to build any applications is stored in a version management repository.
- All source code and data needed to build all infrastructure configurations (infrastructure code and data) is stored in a version management repository.
- All source code and data needed to build all pipeline configurations is stored in a version management repository.
- All results of builds and tests that were used to assess the quality of applications, infrastructure configurations, and pipeline configurations are stored in a version management repository.
- All tests and test scripts needed to verify applications, infrastructure configurations, and pipeline configurations are stored in a version management repository.
- All executable images for release candidate versions of applications, infrastructures, and pipelines are stored in an artifact repository.

- All changes to code and data are tagged with metadata to assist in queries and searches.
- A multimaster version management repository capability keeps track of versions of repos for applications, pipelines, and environment configurations and data as needed to auto-revert the entire environment when needed to roll back to a prior combination of application version, pipeline version, and environment version.
- A dashboard and alert system keep the health of all the version management repos visible always.
- A disaster detection and recovery capability are constantly monitoring for serious failures in the version management system and invoke recovery procedures when detected.
- At least one offline backup of all repositories is kept and available to restore the repositories on demand.
- At least one replica of all the version management repositories are continuously updated.
- The schedule and frequency of saved versions is deterministic and automated. For example, it is typical for every deployed release version to be backed-up for at least three releases, every release candidate including those not released to be backed up for at least two releases, and every development build to be backed up for at least two releases.
- Replication and backup processes are verified periodically—typically at least once per deployed release.
- Access to repositories are controlled with a role-based access management system.

With a well-engineered version management system in place, one of the first things to address is security. Each version of application, pipeline, and infrastructure have possible security vulnerabilities that, left unsecured, could become unwelcome headline news. In the next chapter, a blueprint and engineering practices for continuous security are explained.

8

Continuous Security (a.k.a. DevSecOps)

IT security strategies and tools are at the top of the CIO priority list. Consequences of security attacks can include the following:

- Loss of sensitive or proprietary information
- Disruption to regular operations
- Financial losses relating to restoring systems
- Harm to an organization's reputation

In the context of this book, Continuous Security encompasses both DevSecOps and Rugged DevOps. **DevSecOps** pertains to the culture aspects of "security as code," while Rugged DevOps has to do with engineering security into design and deployment processes. Both have the tenet that security is everyone's responsibility and that DevOps provides opportunities to improve security before deployment.

Why Is Continuous Security Important to Engineering DevOps?

DevOps compliance is a top concern of IT leaders, but information security is seen as an inhibitor to DevOps agility. Security infrastructure has lagged in its ability to become "software defined" and programmable, making it difficult to integrate security controls into DevOps-style workflows in an automated, transparent way. Modern applications are largely "assembled," not developed, and developers often download and use known vulnerable open-source components and frameworks.

How Does Continuous Security Work with DevOps Engineering?

As indicated in **Figure 21—Continuous Security Engineering Blueprint**, the DevOps Continuous Security Blueprint aims to move the organization to a better security posture. Each security flaw is carefully identified and is fixed one at a time to close the most urgent security gaps. DevSecOps identifies the most vulnerable concerns ahead of time and identifies how to avoid or move away from these bad positions.

Without proper consideration given to security engineering practices, the continuous delivery of software changes facilitated by DevOps is risky. On the other hand, DevOps provides an opportunity to reduce security risks if security is integrated into the continuous delivery pipeline according to engineering practices. The following are examples of the *Nine Pillars of DevOps engineering practices for Continuous Security.*[RW56]

Leadership—Leaders need to understand and sponsor a clear vision for security.

- Leadership demonstrates a vision for organizational direction, team direction, and three-year horizon, including security practices.
- Leaders intellectually stimulate the team status quo by encouraging new questions and questioning the basic assumptions about the work, including security practices.
- Leaders promote personal recognition by commending teams for better-than-average work, acknowledging improvements in the quality of work, and personally complimenting individuals' outstanding work, including security practices.

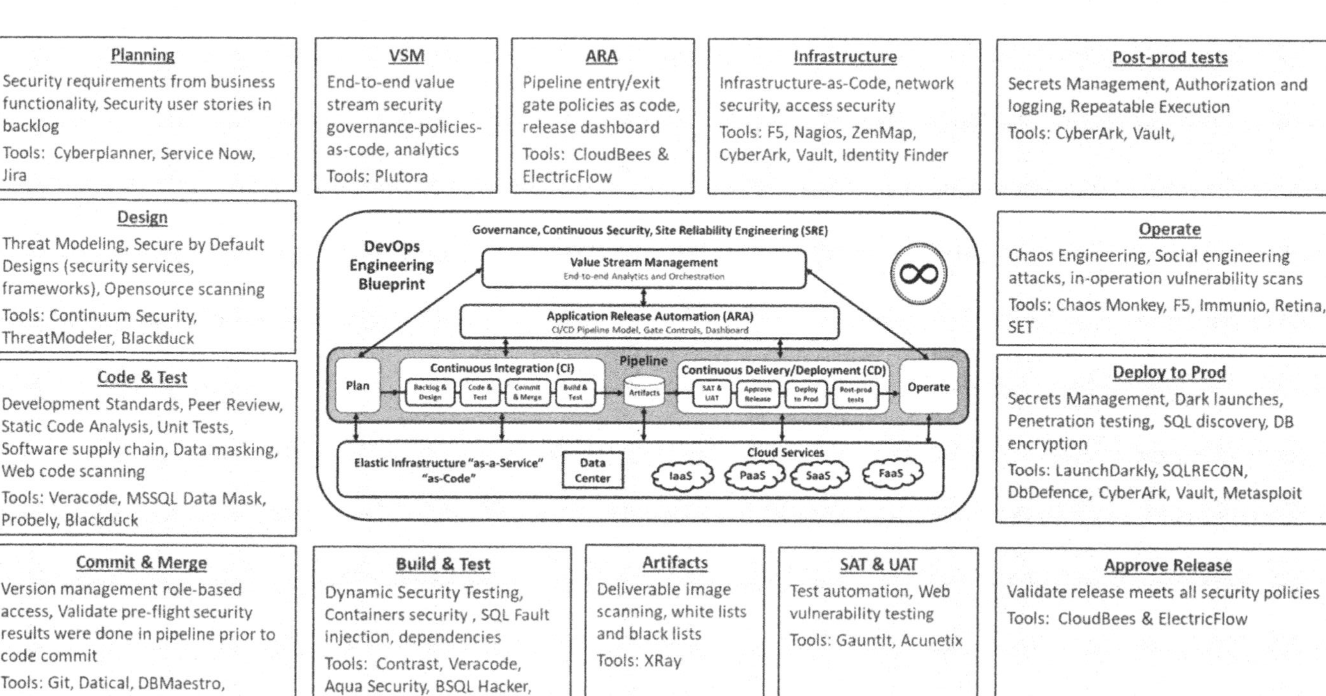

Figure 21—Continuous Security Engineering Blueprint

Collaborative Culture—Culture in organizations that work well with DevOps have a collaborative continuous security mindset.

- The culture encourages cross-functional collaboration and shared responsibilities and avoids silos between developers, operations, project management, quality assurance, and security.
- The DevOps system (toolchain) is created by an expert team and reviewed by a coalition of stakeholders, including security teams.
- Changes to end-to-end DevOps workflows are led by an expert team and reviewed by a coalition of stakeholders, including security.
- DevOps culture empowers and trains team members to take personal responsibility for security, compliance, and privacy obligations.
- Security engineers/architects are involved in the design for modular components and consulted when security patterns change within modules.

Design for DevOps—Designing software for DevOps at speed requires application designers to master the engineering practices for continuous security.

- Software source code changes are pre-checked with static analysis tools prior to commit to the integration branch. This assures that the modified source code does not introduce critical software faults and security vulnerabilities such as memory leaks, uninitialized variables, array-boundary problems, and SQL injection.
- Software component analysis scans third-party components for known security vulnerabilities and identifies risk during the build process.
- Security frameworks for technology stacks that are used (such as Apache Shiro or Spring Security) are documented and shared with developers for their respective technology stacks.
- Software code changes peer code reviews include checks for defensive coding and security vulnerabilities prior to committing code to the integration/trunk branch.

- Common security components such as identity, authorization, key management, audit/log, cryptography, and protocols are maintained, published, readily available, and used within module development.

Continuous Integration—*Continuous integration (CI)* within organizations that have multiple teams working concurrently on a project and different code bases is challenging. During the integration stage, it is critical to assess the application and understand the impact of code changes from a security point of view.

- A software version management system is used to manage versions of all changes to source code, executable images, and tools used to create and test the software.
- Incremental static analysis pre-commit and commit checks are wired into CI to catch common mistakes and anti-patterns quickly by only scanning the code that was changed. These checks identify security vulnerabilities through control flow and data flow analysis, pattern analysis and other techniques. These techniques find security-related issues such as mistakes in using crypto functions, configuration errors, and potential injection vulnerabilities.
- Binary artifacts are digitally signed and stored in secure repositories.
- Changes to security patterns used within software sources such as session management, authentication, authorization, and encryption code trigger a notification or pull request to security engineers.

Continuous Testing—Continuous testing (CT) with DevOps has significant advantages for continuous security when engineering practices are followed.

- New security tests that are necessary to test a software change are created together with the code and integrated into the trunk branch at the same time the code is. The new tests are then used to test the code after integration.

- Security tests for each DevOps pipeline stage are automated and may be selected automatically according predefined criteria.
- Release regression tests include security tests. At least 85% of the security regression tests are fully automated, and the remaining are auto-assisted if portions must be performed manually.
- Test results that indicate possible security concerns are tagged for security analysis.
- Dynamic or interactive application security tests exercise the application for security vulnerabilities. Results of these tests are delivered to developers through tools and feedback loops native to their organization.
- If containers are used, image repositories are scanned for images with known vulnerabilities, hash checks for image drifts, and run-time checks for vulnerabilities during image deploys.
- Attack patterns, abuse cases, and tests are built for application module profiles.
- Unit, functional, and integration tests around—and especially outside—boundary conditions are run during CT. Tests include error handling, exception handling, and logic and negative tests.
- Automated security attack testing to include the OWASP Top 10[RW75] integrated into automated testing.

Continuous Monitoring—*Continuous monitoring (CM)* of security considers the dynamic nature of artifacts and infrastructure and a proliferation of objects and services to secure.

- Metrics and thresholds are automatically gathered, calculated, and made visible to anyone on the team that subscribes to them. Example security metrics include number of security defects identified in pre-production, percent of code coverage from security testing, number of failed builds due to security checks, mean time to detection, and mean-time-to-resolution
- Vulnerability information in consolidated to provide a comprehensive view into vulnerability risks and remediation across tools, pipelines, and apps—and over time.

- Metrics and events from production security controls such as WAF and RASP are used to improve security testing.
- Insight into security threats and events are shared and visible across DevOps teams to enable "attack-driven defense" methodologies.

Continuous Security—*Continuous security (CS)* is itself a distinct pillar with standalone engineering practices that cross all the other pillars.

- All information security platforms that are in use expose full functionality via APIs.
- Immutable infrastructure mindsets are adopted to ensure production systems are locked down.
- Security controls are automated so as not to impede DevOps agility.
- Security tools are integrated into the CI/CD pipeline.
- Source code for key intellectual property on build or test machines is only accessible by trusted users with credentials. Build and test scripts do not contain credentials to any system that has intellectual property.
- External penetration tests (done out of band) scheduled either periodically or on a regular cadence are used to perform deep-dive analysis.
- Telemetry from production security controls such as WAF and RASP are delivered back to development teams to inform application updates.
- Accurate inventory of all software packages and version information is documented via infrastructure as code. Automated detection is used to identify whether any of the packages have known *Common Vulnerabilities and Exposures (CVEs)* associated and define specific remediation actions.

Elastic Infrastructure—Elastic infrastructure environments offer advantages compared to legacy or traditional static infrastructure. However, elastic infrastructures need to follow engineering practices for continuous security because flexible infrastructures offer a broader range of attack surfaces.

- Configuration code includes automated checks, including ensuring unnecessary services are disabled and only ports that need to be open are open and permissions on files, audits, and logging policies are enforced. Development tools are not installed on production.
- Security-approved OS, software versions, and frameworks are used to compose required infrastructure. Security-related controls such as ACLs and FIM are defined as a part of infrastructure where applicable.
- A least-privilege model is enforced for processes running on shared infrastructure.
- Smaller clusters are used to reduce complexity between teams.
- Service provider partners security controls are validated to ensure they meet business requirements in their domains of the shared security model.
- IaaS or PaaS service provider security controls are validated to ensure that they meet business requirements in their domains of the shared security model.

Continuous Delivery/Deployment—While DevOps enables new feature deliveries to users quickly, to minimize risk during deployment, the following engineering practices are important:

- Release-to-production decisions are determined according to predetermined metrics, which include security metrics.
- A whitelist policy for application segmentation is enforced during deployment for each environment, especially production.
- Continuous deployment processes trigger run-time security and compliance checks, including ensuring unnecessary services are disabled and only ports that need to be open are open and permissions on files, audits, and logging policies are enforced. Verify development tools are not installed on production.
- All secrets used for deployment are vaulted and retrieved programmatically during run-time or initialization of the continuous delivery process.

Implementing Continuous Security

The following are recommended steps for implementing continuous security:

- Empower and train developers to take personal responsibility for security.
- Embrace security assurance automation and security monitoring.
- Require all information security platforms to expose full functionality via APIs for automatability.
- Use proven version control practices and tools for all application software, scripts, templates, and blueprints used in DevOps environments.
- Adopt immutable infrastructure mindset where production systems are locked down.
- Security platform capabilities such as identity and access management, firewalling, vulnerability scanning, and application security testing need to be exposed programmatically.
- The integration and automation of these security controls are enabled throughout the DevOps life cycle in automated toolchains.
- Information security teams can then set policies, which can then be applied programmatically based on the type of workload.

In this chapter, it was explained how DevOps continuous delivery without due consideration to continuous security is ill-advised. Engineering practices that integrate security throughout the Three Dimensions (People, Process, and Technology), applications, pipelines, and infrastructures are critical to create a defense against security vulnerabilities. Otherwise security problems can be amplified and accelerated as the release cadence increases. One of the key engineering practices that organizations can use to ensure that security is integrated into value streams and pipelines is the use of a service catalog. This is explained in the next chapter.

9

Service Catalogs Facilitate DevOps Engineering

A DevOps service catalog is an organized and curated collection of business- and IT-related DevOps services that can be performed by, for, or within an enterprise. DevOps Service catalogs act as knowledge management tools for the employees and consultants of an enterprise, allowing them to route their requests for and about DevOps services and related topics to the subject matter experts who own, are accountable for, and operate them. Each service within DevOps service catalog is very repeatable and has controlled inputs, processes, and outputs. When engineered properly, this approach ensures that the application, pipeline, and infrastructure choices that are available from the service catalog includes tools that are needed to measure and calibrate the performance of those choices. **Figure 22—DevOps Service Catalog Examples** shows two example commercial service catalogs for DevOps.

Why Is the Service Catalog Important to DevOps Engineering?

A DevOps service catalog allows organizations to create, centrally manage, and execute applications, toolchains, and infrastructures. For example, a toolchain instance (or collection of toolchain instances) may map to a type of application within the enterprise. A service catalog represents information about all applications, pipelines, and infrastructures that are available and operated within an enterprise.

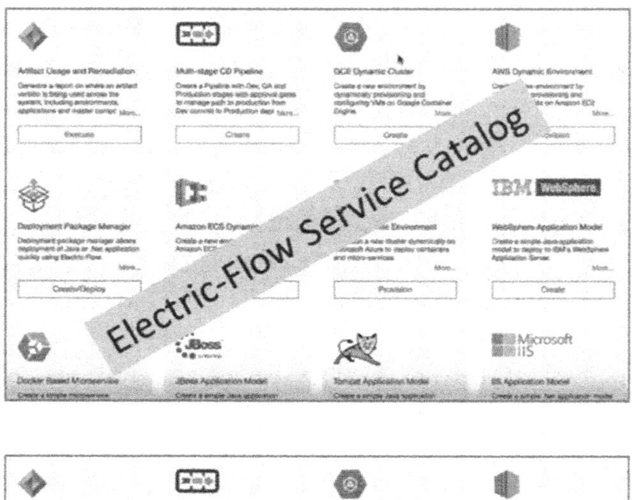

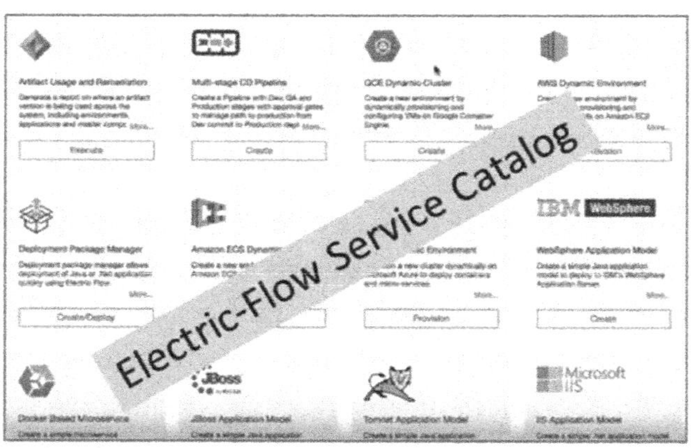

Figure 22—DevOps Service Catalog Examples

Implementing DevOps in large organizations is difficult. Each application can potentially have multiple environments each with multiple and varied *CI/CD pipelines*, tool stacks, metrics, and workflows. Calcified silos result from the inherent legal, geographical, and financial constraints. This makes them culturally resistant to change and less agile. DevOps service catalogs can play a key role in resolving increasing expectations from the business for capabilities like self-service and on-demand provisioning of resources and growing impatience with trouble-ticket culture. DevOps service catalogs can also support the desire to experiment with rapidly evolving productive platforms (e.g., cloud-based) that can accelerate the time to market while controlling spending and technology choices and enforcing corporate policies.

How Are DevOps Service Catalogs Engineered for DevOps?

Service catalogs allow application development teams to onboard their applications and microservices easily with pre-defined templates and pre-approved technologies for typical projects.

Metrics-based goal tracking is the foundation for IT performance. As indicated by **Figure 23—Continuous Feedback Service Catalog Metrics**, tracking performance of applications, databases, continuous testing, releases, infrastructure, and continuous security for a portfolio of rapidly changing products is not feasible without a consistent feedback strategy. A DevOps service catalog is used to ensure *Continuous Feedback* monitoring tools are part of DevOps pipelines and that *Continuous Feedback* monitoring tool choices are consistent, so performance data is made visible across the portfolio of applications.

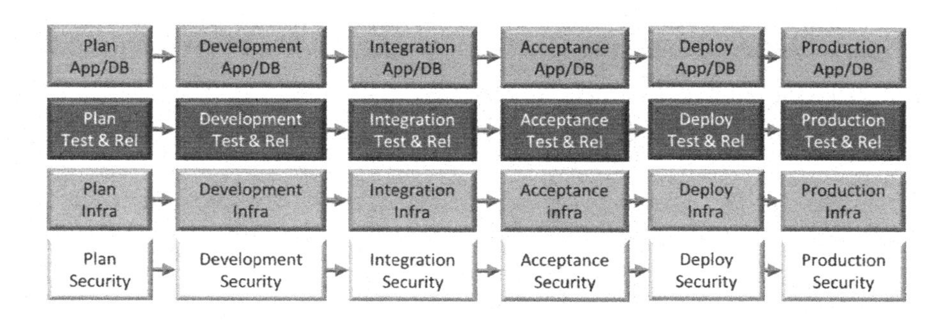

Figure 23—Continuous Feedback Service Catalog Metrics

As indicated by **Figure 24—Continuous Feedback Monitoring Framework**, monitoring tools at each layer feed logs and event data to VSM tool to provide an aggregated view of feedback data.

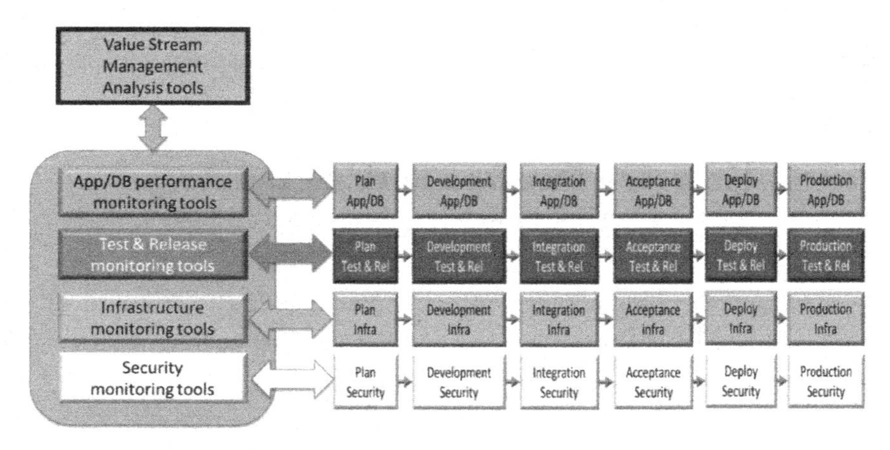

Figure 24—Continuous Feedback Monitoring Framework

The Third Way of DevOps prescribes continual experimentation, which requires taking risks, learning from success and failure, and understanding that repetition and practice are the prerequisites of mastery. As indicated in **Figure 25—Flexible Standardization**, with *The Third Way*, IT must have controlled flexibility to at once ensure standards for toolchains, application stacks, and environments that are used for DevOps and SRE while also facilitating experimentation and changes requested by users. The DevOps service catalog approach needs to be sophisticated able enough to do both. **Figure 26—Flexible Service Catalog Management Workflow** illustrates a DevOps service catalog protocol that offers controlled flexibility.

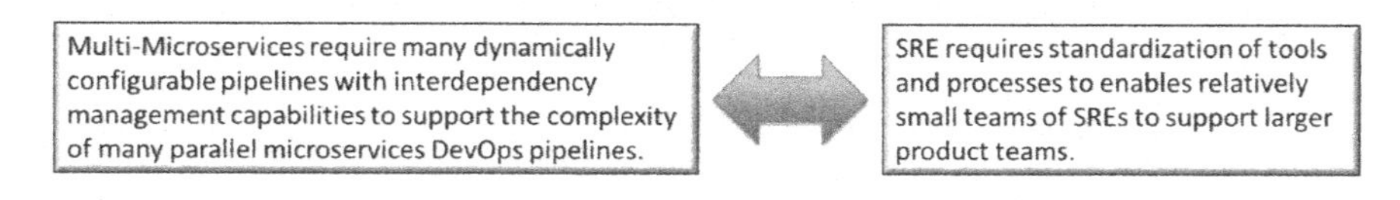

Figure 25—Flexible Standardization

The following are recommended engineering practices for a DevOps service catalog are broken into two categories: DevOps Capabilities Creation and DevOps Capabilities Management.

Engineering practices for DevOps Capabilities Creation are detailed as follows:

- Users can author a toolchain based on a *Domain-Specific Language (DSL)* to ensure that only "approved" tools are included in the toolchain.
- Deployment resources (cloud provider, resources) that need to be provisioned are included.
- Once the toolchain for a given application type has been authored and made available within the DevOps service catalog, users can self-service provision an application via a central UI Portal.
- Users and user groups may customize a custom portfolio for the DevOps service catalog.
- The DevOps service catalog DSL is extensible to include the new tools and services being made available at a rapid clip.
- The DevOps service catalog tool provides the ability to import pre-existing templates from a library of available templates.
- The DevOps service catalog supports a quick and easy way to update the offerings on demand with approval capability.

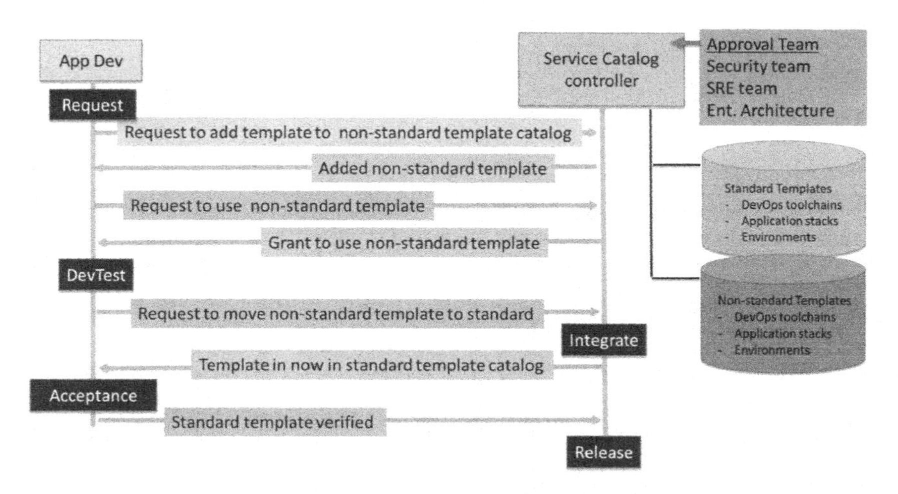

Figure 26—Flexible Service Catalog Management Workflow

Engineering practices for DevOps Capabilities Management are detailed as follows:

- With *Role-Based Access Control (RBAC)*, an end-user can access only the resources that they (or their team) provisioned via the DevOps service catalog—they do not have permission to see resources provisioned by other teams.
- Toolchains include a workflow of steps for approval, escalation if milestones are not met, and notification upon successful completion. This workflow capability is a way to bridge application process guidance and operational process guidance.
- Users can access the status of their submitted DevOps service catalog requests via the UI portal.
- Users can access charge-back information via the UI portal of the DevOps service catalog.
- Control over the resources that are included in the DevOps service catalog is strictly enforced.
- The DevOps service catalog tool supports creating and editing policies. For example, "Enforce an Enterprise-specific naming convention on all provisioned resources."
- Pipeline tools available in the DevOps service catalog include *Continuous Feedback* monitoring tools needed to support data collection for *Continuous Feedback* metrics.
- DevOps service catalog items are access controlled so that only permitted users can change certain items.
- The evolution of the DevOps service catalog itself follows a DevOps approach in which new versions are controlled in a version management database; releases of new versions of the DevOps service catalog follow a controlled and measured pipeline and can be deployed or rolled back quickly.
- The DevOps service catalog allows users to choose to compose a pipeline from a pre-approved selection of standard toolchains, application stacks, and environments.
- Dev and SRE teams may request changes to the toolchain templates, application stacks, and environments lists in the DevOps service catalog for experimentation purposes.
- Records are kept of DevOps service catalog change requests.

- Rules are determined for approving updated or new standard toolchain templates, application stacks, and environments to the DevOps service catalog. For example, SREs must approve any changes that may impact SRE reliability or capacity management processes.

What Is Needed to Engineer a DevOps Service Catalog?

A successful approach to implement a DevOps service catalog includes the following:

- Develop a unified DevOps model that can deal with a large and diverse application portfolio in a consistent manner. This includes consistent monitoring, incident handling, and logging.
- Implement an approach that bridges the application-focused frameworks such as Agile with more operations-centric frameworks like ITIL.
- Break down silos and adopt DevOps without embarking on a "boil the ocean" culture change or completely redrawing the organizational boundaries.
- Extend private data center infrastructures to hybrid and multi-cloud environments so application teams may take advantage of the elasticity and scale offered by the public cloud.
- Start with a small team with a first mover "model" application. Set up clear metrics and automating error-prone and tedious tasks.
- Connect "islands of automation" into a toolchain and bring consistency to aspects like provisioning and deployment.
- Use DSLs (domain specific language; e.g., Azure ARM Templates, Chef Recipes, and AWS Cloud Formation) to bring consistency to the DevOps toolchain. DSL is used to define the order of the tools in the chain, as well as the expression of policies, such as security and configuration. Treating each instance of a toolchain as code has many benefits, including the ability to repeat, the ability to version, and the ability to execute.

- Select tools to implement the service catalog (e.g., AIS Service Catalog, Service Now, etc.)
- Document a DevOps service catalog publishing governance guideline.
- Provide DevOps service catalog development support so the development team can contribute to new and improved service catalog content.

In this chapter, it was explained how Service Catalogs play a useful role in ensuring the choices for applications, pipelines, and infrastructures have built-in tools needed to meet good engineering practices. A well-engineered service catalog goes a long way to supporting governance. The catalog presents a limited number of choices, which helps to contain the possible combinations that need to be governed.

10

DevOps Governance Engineering

IT *governance* is defined as the processes that ensure the effective and efficient use of IT in enabling an organization to achieve its goals. With DevOps, the rate of request for change is continuously increasing, and the cadence of governance decisions needs to respond to this. Uncontrolled proliferation of solutions leads to greater financial cost, a likely divergence in process, and potentially a loss of the feedback and process control that is critical to successful DevOps.

DevOps relies on practitioner empowerment, while DevOps governance *processes* rely on central control.

Why Is Governance Engineering Important for DevOps Engineering?

Proper DevOps *governance* provides the following benefits:

- Taxonomy for communicating governance policies
- Statutory and regulatory compliance (evidence for audits)
- Reinforce enterprise culture, ethics, and behaviors
- Sets direction for use of technology
- Enforce standard operating procedures
- Risk management
- Cost management

- Visibility and process control
- Performance management
- Protect confidentiality
- Facilitates trust with stakeholders

How Is Governance Engineered for DevOps?

DevOps governance is a very different approach than traditional IT governance. It requires rethinking how work is funded and the Lines of Defense framework as follows:

- First Line—Who Owns the Risk—Individual developers and engineers
- Second Line—Who Sets Policy and Monitors the Risk—Governance and Risk functions that set policy and monitor risk daily
- Third Line—Independent Assurance—Internal audits that provide independent insurance and report directly to the audit committee or the board
- Fourth Line—External Partners—Auditors and regulators who must be brought into the conversation and given full transparency into development processes and risk management

A sound governance strategy should cover organizational, financial, and operational requirements. The process must be designed to be repeatable, as the DevOps journey is not just a one-off deal but an ongoing process to facilitate continuous improvement.

Organizational—Define your business services and the organizational roles to deliver those services and break down the traditional silos between development and operations. Then create SLAs to guarantee those as required to support your business services and applications.

Financial—Define your budget for research and development, business support, operational support, network, security (cyber and physical) systems, and IT operations. Then wrap up the process by creating procurement methodologies that will support financial governance.

Operational—Define the processes, procedures, and policies that govern how you will operate and who will operate critical functions. This requires aligning your business goals across your entire organization to ensure that continuous integration and development are achieved.

Figure 27—DevOps Governance Engineering Blueprint illustrates a DevOps governance engineering blueprint that provides for control with practitioner participation.

For *The First Way of DevOps, Continuous Flow* control gates use control logic at the boundary of each pipeline stage to control flow. Example of control parameters for *Continuous Flow* are listed as follows:

- Incident and issues management; thresholds for work items, incidents, and issues. For example, a threshold could be that no priority zero bugs exist.
- Seek approvals outside automated pipelines. For example, approvals from legal approval departments or auditors could be a requirement.
- Quality validation. Query metrics from tests on the build artifacts such as pass rate or code coverage and deploy only if they are within required thresholds.
- Security scan on artifacts. Ensure security scans such as anti-virus checking, code signing, and policy checking for build artifacts have completed.
- User experience relative to baseline. Ensure the user experience hasn't regressed from the baseline state.
- Change management. Wait until change management procedures are complete before deployment.
- Infrastructure health. Execute monitoring and validate the infrastructure against compliance rules after deployment.

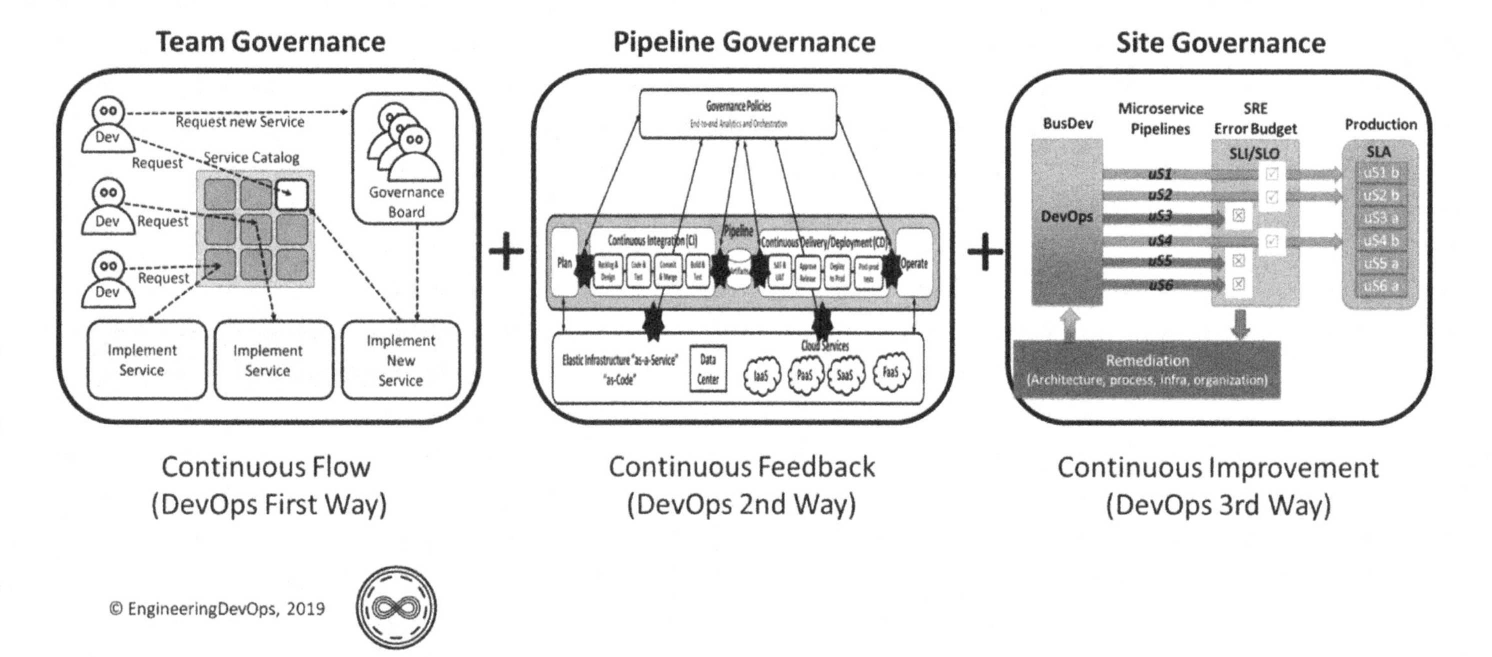

Figure 27—DevOps Governance Engineering Blueprint

The Second Way of DevOps describes the principles that enable fast and Continuous Feedback from right to left at all stages of the value stream (i.e., pipeline). Second Way DevOps systems are essential as an enterprise scales DevOps in complexity to support many parallel pipelines and deployment targets. Measurement and feedback capabilities embedded in DevOps infrastructures and pipelines help shorten and amplify feedback loops. When working in complex systems, real-time feedback provides the earliest opportunity to detect and correct errors before a catastrophic event develops. **As continuous security, continuous testing, and continuous deployment environments scale up, governance must be upgraded to include control gates between pipeline stages that utilize control parameters from feedback sources.**

The Third Way of DevOps prescribes *Continuous Improvement*, which involves continual experimentation, which requires taking risks, learning from success and failure, and understanding that repetition and practice are the prerequisites of mastery. *Third Way DevOps* systems are essential for an enterprise to accelerate its pace of innovative changes while controlling reliability through SRE systems and practices.

Governance practices for *Third Way DevOps* systems build on governance practices of *The First Way* and *The Second Way DevOps* systems. In *The Third Way*, SREs ensure reliability of Service Level Agreements (SLAs) in production by continuously monitoring *Service Level Indicators (SLIs)* during the pipeline to ensure candidate releases that do not meet *Service Level Objectives (SLOs)* are not released to production. Changes that do not meet SLOs are rolled back for remediation.

The following recommended engineering practices for DevOps governance are broken into three categories: DevOps governance for People, Process, and Technology.

Recommended engineering practices for DevOps governance for people are as follows:

- The Governance Policy Board includes cross-functional representation (e.g., Executive, Legal, Product Owners, Security, Human Resources, Architects, Product Developers, Operations, QA,

Infrastructure, Finance, Product Management, Vendor Management, and Project Management) to ensure various stakeholder requirements are covered by the policies.
- Governance policies are reviewed and updated regularly with Development team inputs to keep them current with the rapid changes required for DevOps.
- Product teams deliver ongoing value, and the business can count on improvements to meet market demand.
- Roles of the users of the pipelines include responsibilities to respond to gate criterion exceptions.

Recommended engineering practices for DevOps governance for process are as follows:

- The scope of individual governance policies is defined for all domains (e.g., geographic regions).
- Resource consumption dashboarding, trend analytics, and optimization scenarios views enable you to see what users are consuming and whether certain components can be reduced if they're not needed.
- Every project (test, automation, DevOps, etc.) must adhere to the defined reference architecture using the security blueprints included in a prescribed catalog of services.
- Code changes are monitored to ensure that a "two sets of eyes" peer review take place.
- Unauthorized change monitoring is used as a detective control. All change events are logged. Governance functions and management review what changes are made and determine if they are authorized.
- Access to environments (e.g., production) is limited specific roles. When needed, access is given through time-limited tokens granted under access approval rules (just-in-time admin).
- Gate criteria are defined for each pipeline stage including thresholds for control parameters such as incident and issues, approvals outside automated pipelines, quality validation, security, user experience, change management, and infrastructure health.

- Workflows are implemented for the pipelines to utilize gate information as part of the promotion of changes from stage to stage.
- Gate criteria, tools, and workflows are maintained.

Recommended engineering practices for DevOps governance for technologies are as follows:

- Governance focus is shifted away from a project focus towards a product focus.
- Developers leverage automation, such as building on every commit, implementing static code analysis on every build, scanning for open-source vulnerability, performing static security scanning, and running automated tests.
- A "clean room model" is adopted, where all product pipelines—whether they are application, test, or infrastructure code—are identified and registered under source control.
- Tools for collect information needed for gate criterion decisions are available in the pipelines.

Tools process gate criteria data that is collected. Gate criteria data collection and processing tools are integrated into the pipeline.

What Is Needed to Engineer Governance for DevOps?

The following steps list phases of activities are recommended for implementing governance in accordance with recommended engineering practices.

Phase 1: Define Engineering Governance

- Define gate criteria for each pipeline stage, including thresholds for control parameters such as incident and issues, approvals outside automated pipelines, quality validation, security, user experience, change management, and infrastructure health.

- Identify tools for collecting information needed for gate criteria decisions.
- Identify tools capable to process the gate criteria data collected.
- Production reliability targets (Error budgets = SLO) are determined for each application and service. SREs and Dev teams determine the targets together. This forms a contract between Dev and Ops by which release decisions are governed.

Phase 2: Implement Engineering Governance

- Integrate the gate criteria data collection and processing tools into the pipeline.
- Implement workflows for the pipelines to utilize gate information as part of the promotion of changes from stage to stage.
- Update roles of the users of the pipelines to include responsibilities and respond to gate criteria exceptions.
- Update the gate criteria, tools, and workflows as needed for continuous improvement.

Phase 3: Operations with Engineering Governance

- Site Reliability Engineering practices provide the means (tools, automation, and workflows) to monitor SLIs and act when SLOs are not met.
- SLOs are reviewed periodically to make sure they make sense, as situations tend to change over time during continuous improvement cycles.
- SLIs data collection and processing tools are integrated into the pipeline.

As indicated in the last phase, Site Reliability engineering practices are recommended to help govern DevOps because SRE provides the metrics and disciplined methodologies consistent with operating a well-engineered DevOps environment. In the next chapter, SRE is explained in more detail.

11

Site Reliability Engineering (SRE)

According to the book *Site Reliability Engineering: How Google Runs Production Systems,*[RWB13] "SRE teams are responsible for the availability, latency, performance, efficiency, change management, monitoring, emergency response, and capacity planning of services. SREs codify rule of engagement and principles for how they interact with their environment." *Site Reliability Engineering (SRE)* compliments DevOps by measuring and achieving reliability of applications and services working on production and DevOps infrastructure in a prescribed manner using error budgets, team relationships brokered by an error budget, Ops-as-code, and the use of reliability to control deployments.

Why Is SRE Important to DevOps Engineering?

SRE emphasizes the following objectives that are important to DevOps engineering:

- Assure availability of applications and services
- Assure error rate and latency meet *Service Level Agreements (SLAs)*
- Enable large-scale systems while controlling risk
- Create operational cost savings by automation and policies
- Improve skills of developers and *SREs*
- Resolve conflicts between Dev and *SRE*
- Improve capacity planning and provisioning

How Does SRE Work with DevOps?

Automation improves problems caused by human fatigue, familiarity/contempt, and inattention to highly repetitive tasks. As a result, both release velocity and safety increase. An outage is an expected part of the process of innovation and an occurrence that Dev and SRE teams manage rather than fear. The SRE role is defined as vital part of the organization. Real-time collaboration and communication solutions like Slack are used extensively in organizations that implement SRE.

As indicated in **Figure 28—Site Reliability Engineering (SRE) Blueprint**, a key tenet of *SRE* is an absolute focus on *Service Level Indicators (SLIs), Service Level Objectives (SLOs),* and *Service Level Agreements (SLAs)* as a contract between Dev and SRE teams. Service metrics such as those shown in **Figure 28—Site Reliability Engineering (SRE) Blueprint** are selected for continuous monitoring and action. An error budget (= 1-availability) internal *Service Level Objective (SLO)* is established for each service. Products are required to work within an error budget. Tactics such as phased rollouts and 1% experiments are used to keep within error budgets. Automation is used to implementing progressive rollouts, detect problems, and implement roll-back changes safely when problems arise, because outages are primarily caused by changes in a live system. The *SRE* team is responsible for capacity planning and provisioning. Regular load testing of the system is used to correlate raw capacity (servers, disks, and so on) to service capacity.

The following recommended engineering practices for Site Reliability Engineering (SRE) are presented in two categories: SRE visibility practices and SRE control practices.

SRE visibility practices are detailed as follows:

- End-to-end release processes are made visible and controlled.
- Real-time status of releases is made visible. This enables proactive decision-making around release scope, schedules, and resources to achieve optimal business objectives.
- A real time dashboard displays the current status of all the releases.

- Team members and automated tools can update their release activities.
- SRE tools accept changes to release schedules and provides visibility into the impact of changes to releases.
- Release dependencies are visible using a tool that automatically generates an impact matrix showing releases, application changes, and their scope.
- Compliance audits are supported by a tool that keeps track of activity (who did what, and when) and supports approval workflows.
- Deployment plans, including task assignments and approval workflows, are captured in a tool.
- Capacity planning is supported with a tool that captures environment usage and provides utilization reports.
- Execution times of deployment steps are captured in a tool to support analysis for continuous improvement.
- Environment configurations are captured in a tool.
- Dependencies between environments are made visible with a tool.

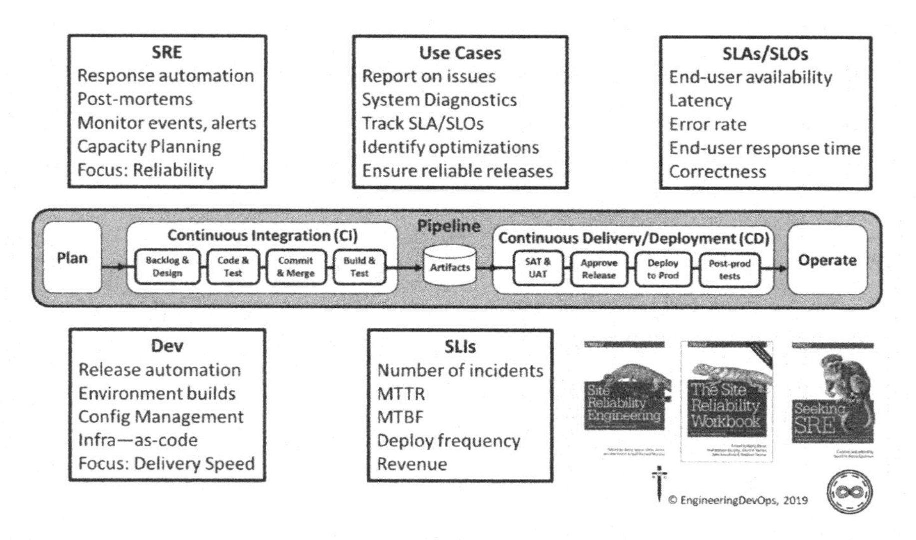

Figure 28—Site Reliability Engineering (SRE) Blueprint

SRE control practices are detailed as follows:

- Dependencies between deployment steps are captured in a tool.
- A tool tracks deployment task assignments, notifies responsible parties of their tasks, and provides real-time status of deployments.
- Issues and incidents occurring during deployment are captured in a tool to support analysis for continuous improvement.
- Shared environment conflicts are managed using a tool that provides visibility into environment availability and supports workflows to request/release environments.
- Booking requests for environments are handled using a self-service tool.
- Patching and maintenance of environments are supported by a tool that indicates when environments are down.
- Environment change requests are supported by a tool that can pull environment configuration data from existing CMDB, discovery, deploy, or ITSM tools.
- Compliance audits are supported by a tool that captures environment changes and approvals.

What Is Needed to Engineer SRE with DevOps?

The following steps are recommended for implementing SRE:

- SRE and Site Reliability Champion roles are defined. "Site Reliability Engineer = Software Engineer + Systems Enthusiast." "I'm going to take the time to automate this right now and stop anyone else from having to do this painful thing."
- SRE teams include software developers with strong operations knowledge or IT operations people with strong software development skills.
- Codify rules of engagement and principles for SRE teams.
- Cap the aggregate Ops work for all SREs.

- Standardization of both tools and processes enable relatively small teams of SREs to support larger product teams.
- Monitor solutions with software interpreting, only notifying humans when they need to act.
- Playbooks record the recommended engineering practices for human response actions.
- Automate progressive rollouts, problem detection, and roll-backs.

As indicated in this chapter, SRE provides visibility and control over functioning DevOps environments. However, disasters can happen that interrupt the operation. The next chapter explains recommended engineering practices to mitigate DevOps disasters.

12

DevOps Disaster Mitigation and Recovery

Well-engineered systems are designed and implemented to anticipate failures and handle failure modes gracefully. There is no such thing as absolute disaster prevention. Unpredictable failures caused by actions from humans, automation, and nature in applications, pipelines, and infrastructures will cause a disaster event that impacts the Nine Pillars of DevOps. The impact of DevOps disasters can range from the embarrassment of a failed demonstration, loss of sales, or total business collapse, such as the Knight Capital case discussed earlier in this book. Disasters will happen even with the best DevOps environments. There is no good time for a disaster, but there are times that are worse than others, such as the major business impact of disaster events at the end of a business quarter, during business valuations, or during the peak business time of an e-commerce website.

Disaster mitigation and recovery strategies minimize the risk and impact of disasters. The good news is that well-engineered DevOps environments are also well-engineered for disasters. DevOps practices covered in this book (including VSM; version management; application release automation; governance; site reliability engineering; continuous security; and CI/CD practices such as monitoring, promoting, and rolling changes to applications, pipelines, and infrastructures in small increments) all help to minimize both the blast radius of disasters and the time to recover from disasters.

Figure 29—DevOps Disaster Mitigation and Recovery Blueprint shows examples of disasters that affect applications, pipelines, and infra-

Nine Pillars of DevOps (Cause)	Application Disaster	Pipeline Disaster	Infrastructure Disaster
Leadership	Example: App crash w/bad release approval Mitigation: Version Management with VSM Recovery: Roll-back to last good App version	Example: Pipeline crash w/bad change approval Mitigation: Version Management, with VSM Recovery: Roll-back to last good Pipeline version	Example: Infra crash w/bad change approval Mitigation: Version Management, with VSM Recovery: Roll-back to last good Infra code version
Collaborative Culture	Example: App crash w/ poor change coordination Mitigation: Dependencies Version Management Recovery: Roll-back to last good App version	Example: Pipeline crash w/ poor change coordination Mitigation: Dependencies Version Management Recovery: Roll-back to last good Pipeline version	Example: Infra crash w/ poor change coordination Mitigation: Dependencies Version Management Recovery: Roll-back to last good infra code version
Design For DevOps	Example: Monolith App crash during operations Mitigation: Redundant deployments Recovery: Switch to last good version of App	Example: CI crash w/application file corruption Mitigation: Redundant image repository Recovery: Switch to last good back-up file	Example: Infra crash w/config code file corruption Mitigation: Redundant image repository Recovery: Switch to last good back-up file
Continuous Integration	Example: App crash w/coding error Mitigation: Static analysis checks Recovery: Roll-back to last good App version	Example: Bad CI automation script breaks CI process Mitigation: Version management Recovery: Roll-back to last good Pipeline version	Example: Infra crash with bad config integration Mitigation: Version management Recovery: Roll-back to last good Infra code version
Continuous Testing	Example: App crash w/test exception Mitigation: Test validation Recovery: Roll-back tests to last good version	Example: Pipeline crash w/test exception Mitigation: Test validation Recovery: Roll-back tests last good Pipeline version	Example: Infra crash w/test exception Mitigation: Test validation Recovery: Roll-back tests last good Infra version
Continuous Monitoring	Example: App crash w/performance exception Mitigation: APM design validation Recovery: Roll-back APM config to good version	Example: Pipeline crash w/performance exception Mitigation: APM design validation for Pipeline Recovery: Roll-back APM config to good version	Example: Infra crash w/performance exception Mitigation: APM design validation for Infra Recovery: Roll-back APM config to good version
Elastic Infrastructure	Example: App crash due to failed cluster Mitigation: Load balancer Recovery: Restrict load and add clusters	Example: Pipeline crash due to failed servers Mitigation: Load balance Recovery: Restrict load and add clusters	Example: Infra crash due to network failure Mitigation: Chaos monkey, alternate networks Recovery: Switch to alternate network
Continuous Security	Example: Cyber attack takes down Application Mitigation: Continuous security best practices Recovery: Recover data and deployments	Example: Hacker takes down Pipeline Mitigation: Continuous security best practices Recovery: Reset and restore pipeline	Example: Hacker takes down Infrastructure Mitigation: Continuous security best practices Recovery: Reset and restore infrastructure
Continuous Delivery	Example: New version of App fails in production Mitigation: Dark launches of App releases Recovery: Roll-back to last good App version	Example: New version of Pipeline fails Mitigation: Dark launches of Pipeline releases Recovery: Roll-back to last good Pipeline version	Example: New version of Infra fails in production Mitigation: Dark launches of Infra code Recovery: Roll-back to last good Infra version

Figure 29—DevOps Disaster Mitigation and Recovery Blueprint

structures caused by failures of one of the Nine Pillars of DevOps and details how to engineer mitigations and recovery strategies for them.

A common mitigation strategy for disasters is to periodically test fault detection and recovery procedures for applications, pipelines, and infrastructures. A successful disaster mitigation strategy pioneered by Netflix is called Chaos Engineering,[RW74] uses a "chaos monkey" that deliberately injects failures into the application, pipeline, or infrastructure to test the infrastructure's ability to detect, tolerate, and recover from failures.

PART III

ENGINEERING APPLICATIONS, PIPELINES, AND INFRASTRUCTURES FOR DEVOPS

"That is the Lady of the Lake, said Merlin; and within that lake is a rock, and therein is as fair a place as any on earth, and richly beseen; and this damosel will come to you anon, and then speak ye fair to her that she will give you that sword."

—**Le Morte d'Arthur**, *BOOK I CHAPTER XXV*

In this part, **Engineering Applications, Pipelines, and Infrastructures Engineered for DevOps**, a comprehensive explanation of recommended engineering practices for the lower levels of the DevOps Engineering Blueprint is presented in six chapters as follows: "DevOps Application Engineering," "CI/CD Pipelines Engineering," "Elastic Infrastructure Engineering," "Continuous Test Engineering," "Continuous Monitoring Engineering," and "Continuous Delivery and Deployment Engineering."

13

DevOps Application Engineering

Did you ever buy a "one-size-fits-all" item, only to discover that you must be a mutant of humankind because when you tried it on it didn't fit? Or maybe the manufacturer forgot to clarify the species that the item was designed for. Maybe it was meant for fairies instead? If so, then you will relate to this section.

The extreme amount of world-wide hype promoting DevOps products and services could suggest that DevOps is good for everybody and applies equally well to every software-based product or service and organization. Here is my experience with that: It's not true! On the other hand, I have heard people say that DevOps does not apply to applications that it absolutely does!

Before you go a step further in your DevOps journey or decide that your application doesn't qualify for DevOps, you should consider carefully how well your application is engineered to fit to become a model for DevOps implementations.

The following are important factors for deciding whether an application is engineered for effective DevOps:

- The application will benefit from faster lead times.
- Leaders over this application are open to collaboration and will be sponsors of change.

- Team players that are associated with the application team (product owners, Dev, QA, Ops, Infra, Sec, PM) are open to collaboration and change.
- The application is currently using or planning to use service-oriented, modular architectures or microservices.[RB14, RB15, RB16]
- There are at least ten people associated with the application including product owners, Dev, QA, Ops, Infra, Sec, and PM. Smaller projects with smaller teams may not show sufficient impact or justify substantive DevOps investment.
- The application is expected to be undergoing changes for more than a year. Less time would not likely yield return on the investment required for DevOps.
- The application represents a good level of business impact and visibility but does not involve an extreme amount of risk for the business.
- The application experiences frequent demands for changes from the business.
- Not all tools used with the application need to be replaced to implement a DevOps toolchain. The costs of changing out all tools can negatively affect the ROI and business case for DevOps transformations.
- Efforts to build, test, or deploy releases of the application are significant and could be reduced significantly by automation.

It is not necessary for an application to perfectly fit all the factors, but the more it fits the more likely your application will benefit from using DevOps and will therefore be worth the cost to implement a well-engineered DevOps. To make it easy for you to evaluate your application, a DevOps Model application scorecard is provided in Appendix E.

!! Key Concept!! DevOps Application Scope

DevOps can be used with all types of software applications and services. However, not all applications and services should be engineered to use DevOps. The factors that determine whether to engineer DevOps for an application have more to do with business factors than technology differences between different types of applications. Applications must be engineered in specific ways for DevOps to operate effectively.

Application Design for DevOps

"The beatings will continue until morale improves." "We must keep testing until the software quality is good enough." "We cannot stop to improve things because we are too busy doing it." "We can't afford the improvements because we need the money to operate the way we do now." Really? Who says so? I hereby recommend that the person saying those things should be fired and replaced with someone who has a more strategic view. Unfortunately, this type of short-term versus long-term thinking permeates many organizations, and the effects creep into designs of software-based products and service. Poor design practices severely limit the potential for improvements provided by DevOps.

As stated in my blog post "Design for DevOps—Recommended Engineering Practices,"[RW20] "Without good design practices, a DevOps implementation has no hope of delivering on its promise of accelerating innovation with high quality and scale." Following DevOps-recommended engineering practices for applications design is important for legacy, greenfield or brownfield products, platform projects, feature enhancements, or repairs. It is important for enterprise three-tier applications or multilayer products. And it is true whether the developers are using Agile, Waterfall, or ITIL processes.

In every application design case, designers face tough challenges and are expected to balance conflicting goals. Designers are expected to rapidly drive down the work backlog yet produce quality products that avoid costly rejections and roll-backs. In addition, there is pressure to increase the percentage of effort spent on creative content over corrective content and do so with limited time and resources. These conflicting goals constitute tall responsibilities. If not supported with recommended engineering practices, the overall design experience is a cauldron of stress and a fast route to burnout.

The following recommended engineering practices improve the design experience and products of design in a way that is streamlined for the DevOps pipeline:

- Designers must thoroughly understand customer use cases.
- Usability, reliability, scaling, availability, testability, and supportability are more important than individual features! Quality over quantity is recommended. Designs that anticipate actual customer usage are the most successful.
- The culture needs to support designers.
- Leaders must support the designers with motivation, mentoring, and training. No designer can be expected to know everything. It's OK for designers to make some mistakes if lessons are learned and improvement quickly follows. Continuous monitoring and quick remediation are examples of good DevOps practices that help minimize the impact of any mistakes.

Design coding practices are critical in the following ways:

- Products are architected to support modular independent packaging, testing and releases. In other words, the product itself is partitioned into modules with minimal dependencies between modules. In this way the modules can be built, tested, and released without requiring the entire product to be built, tested, and released all at once.[RB16]
- Where possible, applications are architected as modular, immutable microservices ready for deployment in cloud infrastructures, in accordance with the tenets of twelve-factor immutable non-monolithic apps, rather than monolithic mutable architectures.[RB14, RB15]
- Software code changes are pre-checked using peer code reviews prior to commit to the integration/trunk branch.

- Software changes are integrated in a private environment together with the most recent integration branch version and tested using functional testing prior to committing the software changes to the integration/trunk branch.
- Developers commit their code changes regularly—at least once per day.

Good DevOps design practices support QA in the following ways:

- Design practices consider and understand the QA process.
- Software code changes are pre-checked with unit tests prior to commit to the integration/trunk branch.
- Software source code changes are pre-checked with static analysis tools prior to commit to the integration branch. Static analysis tools are used to ensure the modified source code does not introduce critical software faults such as memory leaks, uninitialized variables, and array-boundary problems.
- Software code changes are pre-checked with dynamic analysis and regression tests prior to commit to the integration/trunk branch to ensure the software performance has not degraded.

Good DevOps design practices support Ops in the following ways:

- Design practices consider and understand delivery and deployment pipeline processes.
- Software features are tagged with software switches (i.e., feature tags or toggles) during check-in to enable selective feature level testing, promotion, and reverts.
- Automated test cases are checked in to the integration branch at the same time that code changes are checked in. Evidence that the tests passed are included with the check-in.
- Tests are conducted in a pre-flight test environment that is a close facsimile of the production environment.

The following supporting tools are needed to realize design for DevOps practices:

- Elastic infrastructures that can be easily orchestrated, created, and released as needed to support designers' tasks on demand with minimal delay
- Design, code management, monitoring, and test tools readily available and scalable with minimal delay
- Monitoring tools that track application process performance and report the results to designers in easily consumable formats without delay

Applications for which DevOps Does Not Apply

"If you build it, they will come" made for a great movie idea in *Field of Dreams*,[RW21] but the reality is that many DevOps journeys have led to nowhere despite the excellence of the people, process, and technology involved. As indicated earlier in this section, applications that will not benefit from faster lead times, have insufficient costs to justify investment in DevOps, or are too rigid to accept change are not good candidates for DevOps.

One of my clients, together with twenty of their IT managers and senior engineers, were wrapping up a four-week DevOps assessment project. During the final day in the assessment results read-out meeting, an executive walked into the room and asked, "What application did you assess for DevOps?" The project manager proudly announced the name of the product he had selected for the assessment. The exec turned red and screamed out, "Didn't you know that product is already scheduled to be put on 'end-of-life' status this quarter!" A follow-up action was assigned to the project manager to consider whether or not assessing a more suitable application for DevOps would be more beneficial. Meanwhile I heard the same organization has a critical new application that suffered a ten-month delayed release to market. I can't help but assume that the outcome could have been different if the more suitable application had been assessed instead.

DevOps Applied to Enterprise Apps

An enterprise application assists an organization in solving business problems. Enterprise applications are typically designed to interface or integrate with other enterprise applications used within the organization and to be deployed across a variety of networks (Internet, Intranet, and corporate networks) while meeting strict requirements for security and administration management. Proprietary enterprise applications are usually designed and deployed in-house by specialized IT development teams within the organization. However, an enterprise may outsource some or all the development of the application and bring it back in-house for deployment.

The size and complexity of large enterprise applications, especially older monolithic architected applications, that are not designed in modules are challenging for implementing DevOps. Given the large scale and distributed nature of the people, process, and technology components of enterprise applications, it is advised to engage very experienced DevOps consultants to determine strategic requirements, roadmaps, and plans for DevOps that fully consider all Three Dimensions and Nine Pillars of DevOps before embarking on a major DevOps journey.

Despite the complexity, with the right strategy and plan DevOps can deliver enormous ROI for enterprise applications. *The DevOps Handbook*[RB7] describes customer success stories for enterprise applications. The DevOps Enterprise Summit[RE1] event has hosted many corporations that showcased how they have succeeded with DevOps. Amazon and Adobe are just two of many examples of large corporations that have benefitted enormously from applying DevOps across the enterprise.[RW23]

DevOps Applied to COTS Systems

I often hear that DevOps does not apply to systems that use commercial-off-the-shelf (COTS) software. The idea that DevOps cannot be used with COTS systems is a cop-out. Just because someone else designed and developed the source code is not a sufficient reason to not engineer a DevOps solution.[RW24] Refer to the value stream. Where are the bottlenecks in the chain of activities from the source vendor through integration and deployment in your environment?

There are limitations and constraints to the level of automation allowed, the return on investing in automation, and influence over the development lifecycle that DevOps practices can introduce when dealing with COTS applications.[RW73] COTS applications will typically have predefined and supported interfaces for managing configuration or accessing data. These endpoints may not be accessible or compatible with your existing tool chain or automated change management practices. COTS applications typically come in one of three flavors, namely closed, open, and platform, and each flavor should be evaluated differently.

Closed COTS applications allow little to no customization to functionality and/or interfaces. They typically have predefined management consoles and a published list of commands or APIs that will allow external applications to interact with the COTS applications. A good example of a closed COTS applications is Microsoft Exchange. Key considerations for DevOps with Closed COTS applications are as follows:

- Can configurations be scripted and integrated with existing automation and orchestration tool chains?
- Can configuration scripts be managed and maintain in external version control systems?
- How are updates, patches, and service packs applied and managed? Can these practices be automated?
- What developer tools (APIs, Software Development Kits [SDKs], etc.) are available to extend and/or interact with the application and/or data?
- Are these tools compatible with existing tools and expertise in the enterprise?
- Can these developer tools be integrated within the existing toolchain?
- How frequently are changes made to the base product?

If these applications can be managed by external systems and code developed to automate deployment and configuration and/or integration can be managed through version-controlled pipeline, closed COTS application make good candidates for DevOps and CD. By applying CD principles, COTS binaries are versioned and stored in an artifacts repository, install and configure scripts are managed through version control and IaaS workflows, and applications and integration endpoints are auto-tested through deployment pipelines. This is only made possible when consumers of the COTS application collaborate with operators of the COTS platform to ensure that the solution is aligned with enterprise goals, objectives, and standards related to portfolio management.

Open COTS applications allow significant modifications to functionality, data, and/or interfaces. These platforms typically have rich SDKs, APIs, and embedded developer utilities that enable users and developer to modify all layers of the applications, namely presentation, business logic, and data. Open COTS applications typically have a large, complex footprint consisting of core services, data and/or GUI customizations, embedded applications, etc. Classic examples of open COTS solutions include ERP/CRM platforms (like SAP or Oracle) or portal platforms (like SharePoint, ServiceNow, or WebSphere). The following are key questions when considering whether for DevOps can be applied to an Open COTS application:

- How are customization created through internal design editors, logic builders, or data schema extensions version controlled? (Are these configurations stored in database or code?)
- How are customizations created through internal tools and packaged, installed, and configured in higher-level environments like Stage or Production?
- What does the development life cycle for customization created with internal tools look like? What quality gates are required? How about architectural standards?
- How are custom applications/applets that are embedded in the application packaged, installed, and configured?

- Can these custom applications/applets be run and tested separately from the underlying COTS application?
- What does the development life cycle for customizations created with external tools look like? What quality gates are required? How about architectural standards?
- Does the COTS application provide testing frameworks, mocks, and/or stubs for testing external dependencies?
- What automated test suites are supported by the platform (unit, functional, regression, capacity, etc.)?
- What is the upgrade path for customizations (historically)?

As you can imagine, open COTS applications add layers of complexity. When considering DevOps and CD with open COTS solutions, it is recommended to build multiple pipelines to support each layer of the application in the DEV or CI stage of the value stream. Subsequent stages like Stage and Prod will use a converged pipeline built from known good artifacts of the DEV and CI stages. Before implementing a solution, it is important to study the development lifecycle and practices of the target COTS application before building your toolchain. Understanding how the end-state platform is created and how change is introduced will strongly influence your toolchain and pipeline design.

Platform COTS Applications provide a set of services and tools as well as proprietary runtimes that enable users to build and run custom applications on the platform. These are the most difficult type of applications to integrate into existing DevOps practices and CD toolchains because common practices like version control, testing, and build are done through the platform rather than from external tools. Sample platform COTS applications include Pegasystems and IBM Case Manager. Platform COTS are packaged applications that enable customers to build applications on top of the base platform. As a blend of closed and open systems, all considerations are valid. In most cases, the base platform acts as a closed system, whereas the custom-developed applications on top act like open systems. For this type of application, it is critical to understand what hooks and services are available to support DevOps practices and CD. Many of these platforms are not going to fit within your existing tooling and may not be able to employ the workflows, dashboards, and reporting you have grown accustomed to from your current tool chain.

DevOps Applied to Manufactured Software Embedded Products

In the late 1970s (yes, forty years ago!) as a young engineer, I was given an assignment to figure out how to accelerate the software development and delivery cycles for an innovative packet switch called SL-10, (predecessor to the more recent model called Data Packet Node [DPN]) manufactured by Northern Telecom. Each ***SL-10*** "node" consisted of millions of lines of original embedded software source code and custom hardware configured with many interconnected processors and protocols interfacing with other nodes, host devices, and terminal devices.

Given the complexity of the system and lack of automated build, test, and deployment tools, it was taking more than twelve months to generate each release of the system that typically had lots of bugs even after the release. With knowing much about DevOps (the word "DevOps" wasn't invented until twenty-nine years later), we proceeded to implement a continuous delivery system that included automated build, test, and delivery tools and workflows.[RR5, RR6, RR7]

The result was remarkable. Lead time for releases to manufacturing were reduced more than ten times from twelve months to one month, and quality, measured as defects found by customers per release, also reduced tenfold. This dramatic shift in performance and capabilities made it possible for the engineering and manufacturing teams to accelerate the rate of innovation and trample the competition. The SL-10 product, and its rebranded version called DPN, quickly dominated the connection-oriented packet switch market, and it became the pride of the Nortel corporation and the Canadian nation. Even cooler, I got a nifty president's pin and promotion and paved the path for my career that led to writing this book! If this real-life example is not sufficient to demonstrate that DevOps can be applied to manufactured embedded software products, then I don't know what is.

There have been many other examples published that show how DevOps practices applied to embedded manufactured systems have reaped similar benefits. A notable example was published in 2013 by Gary Gruver in *A Practical Approach to Large-Scale Agile Development: How HP Transformed LaserJet FutureSmart Firmware.*[RB12]

That is not to say that DevOps applied to manufactured software embedded systems is the same as applying DevOps to Enterprise applications or COTS systems. A paper published by Electric Cloud on their blog site, "Five Aspects that Make Continuous Delivery for Embedded Different,"[RW25] does a good job of explaining some of the key practices for DevOps used with embedded systems. These are summarized as follows:

- Retrofitting the huge code base typically found in legacy embedded software systems into a DevOps model is not easy and is likely to be expensive, but it is almost guaranteed to be a worthwhile effort, especially if you have a longer-term vision and intend for your product to stay competitive in the future.
- Developers of embedded software need access to massive amounts of computer resources for testing their real-time embedded code on powerful hardware simulators.
- Developers and testers need access to physical product hardware to run their binaries for system tests that cannot be run on simulators alone. *Lab-as-a-Service (LaaS)* systems that orchestrate testing topologies of real devices can improve sharing of critical resources. LaaS systems provide a test framework that can be integrated through APIs into a continuous delivery pipeline. I know this because I was the inventor of one of the first of these LaaS products, called Lab Manager, at a company that I founded, EdenTree Technologies.[RW26]
- Long build and test processes may need to be re-engineered to reduce serialized lead times into quicker build times made possible by scaling build and test resources.

- Long compliance and conformance test processes for embedded systems need to be accelerated with automation and resource scaling.
- Focus on continuous delivery rather than continuous deployments in which shippable product versions are frequently delivered to manufacturing for verification, but the frequency of deployments to end customers is determined by other business factors.

DevOps Applied to Software Services

I hear people ask, "Does DevOps apply to software services as well as it does products?" These are usually people who have only worked in product-oriented organizations, such as companies that manufacture things that have embedded software. Examples are companies that make things with embedded software and for which DevOps can be used are manufacturers of cars, medical devices, robots, pet trackers, smart home devices, smart toys, and other "things." In addition, companies that "manufacture" software products that are distributed on media or downloaded to be installed on a server or a personal device can also use DevOps.

Ironically, I often hear people ask the opposite question: "Does DevOps apply to products with embedded software as well as it does to services?" These are usually people who have only worked organizations in which software is consumed as a service over an intranet or internet for a price or for free. Examples of these organizations are IT departments (even the IT departments of manufacturers!); government agencies; and almost every enterprise such as banks, insurance companies, hospitals, and financial trading companies.

I assume you already know the answer to these two questions if you read the earlier subsections on embedded software and Enterprise software. The answer to both is clearly YES!

Cloud services are doubly interesting to DevOps because not only are Cloud services used by DevOps as a preferred infrastructure, as described in the next chapter, but Cloud Services also USE DevOps in

the process of creating and deploying software for the implementation of cloud components.

While we are on the topic of DevOps and services, I want to help clarify some confusion regarding whether "newer" DevOps renders "older" *IT Service Management (ITSM)* obsolete or less important. This confusion exists because *ITSM* has often been regarded as a gatekeeper run by Ops to prevent Dev from getting out of control. While some organizations may have in fact implemented *ITSM* that way, this is not the purpose of *ITSM*. "DevOps tends to focus on the software delivery lifecycle and on building and delivering software," said Groll, president of the ITSM Academy. For its part, ITSM tends to have a broader focus; it's really looking at all aspects of customer engagement, from fulfilling standard services to ongoing support and operations activities. So ITSM is "almost like the left and right ends of DevOps."[RW28]

Here are some recommended engineering practices for using ITSM with DevOps:

- Training to help DevOps professionals gain an understanding of ITSM, and vice versa, will help both teams find points of integration—not just in technology but in process as well.
- Find ways to automate service change approvals implementation of known, good changes to allow for agility, iteration, and innovation. At the same time, hold back high-risk, high-impact changes that might need more than just an individual contributor to sign off on before they make it in front of a customer.
- Self-service requests, automation flows, and IT orchestration are now the front door to IT service management actions. These can and should be happening in the background, automagically created, maintained, and measured by our DevOps tools.
- Embrace Agile service management aims to take Agile values and apply Scrum-like development practices to process design and improvement. Aim to make more frequent incremental changes rather than design and implement a total end-to-end process in the traditional way.

Five Levels of Application Maturity

Why does it make sense that younger applications are more mature than older applications? DevOps works best with incremental changes rather than large complex changes that are time-consuming to change, build, test, and deploy. DevOps also works best in *ephemeral* cloud computing infrastructures that can be scaled up and down quickly depending on short-term workload demands of supporting build, test, and deployment requirements. While existing applications will be modernized to take advantage of the cloud, Greenfield applications should be designed to exploit recent advancements in cloud-based platforms and application delivery. The following paragraphs describe the characteristics for five levels of Application Maturity.

Spaghetti applications are older, monolithic legacy applications designs that tend to be large and complex, like a big messy bowl of spaghetti. This complexity occurs partly because these older apps accumulate a massive amount of code and patches over a long time and partly because the architectural approaches at the time, they were conceived did not have modular, cloud-oriented requirements in mind. Spaghetti applications are least able to take advantage of cloud infrastructures.

Modular applications emphasize architecture designs that separate components of an application into functionally independent, interchangeable modules, such that each contains everything necessary to execute only one aspect of the desired functionality. A module interface expresses the elements that are provided and required by the module. The elements defined in the interface are detectable by other modules.

Modules often use a design style known as Service-Oriented Architecture (SOA),[RW31] where services are self-contained black boxes that provide capabilities to the other application modules as a service through a network protocol. SOA may be confused with Microservices architectures.[RW31] There are key differences between SOA and Microservices architectures of importance for DevOps. Microservices are preferred over SOA architectures. Microservices are more fine-grained and bounded in what they can do. SOA uses an Enterprise Service Bus for communication, whereas Microservices use more lightweight stateless protocols and APIs managed as a product. Changes to SOA modules usually require a new application release while a change to a microservice usually only requires the microservice to be replaced.

Migration of modular applications to the cloud may require rework to take advantage of cloud features such as auto-scaling.

Cloud-ready applications use **The Twelve-Factor Application** modular design methodology.[RW21] This is a more advanced modular design methodology that emphasizes declarative formats for automation, clean contracts with the underlying operating system, and portability between execution environments. Deployment on modern cloud platforms obviates the need for servers and systems administration, minimizing divergence between development and production and enabling continuous deployment for maximum agility. These characteristics and capability enable the application to scale up without significant changes to tooling, architecture, or development practices. Cloud-ready applications can be easily mapped to *Virtual Machines (VMs)* available through cloud *infrastructure services (IaaS)* for VMs based on *VMware* vSphere or *Microsoft Hyper-V. Cloud-ready apps* do not require re-architecting or refactoring. They offer least-effort migration to the cloud with a lower availability rate. However, workloads such as legacy databases, third party applications, and homegrown business applications that are typically not designed for clustering or high availability suffer from a single point of failure. *Cloud-ready apps* are expensive to maintain. This is primarily due to the way they consume computing, storage, and network resources, even when idle. Each component of an app runs in a dedicated *VM* with no scope for optimization. Upgrading or patching results in downtime makes the maintenance complex and expensive. Ultimately, *cloud-ready apps* derive less value from the cloud than Cloud-Optimized or *Cloud-Native Applications.*

Cloud-optimized applications take advantage of managed services offered by cloud providers. For example, instead of running a database in a

VM, they consume *DB-as-a-Service*. *Cloud-optimized apps* typically target *Platform as a Service (PaaS)* delivery model of the cloud. They are portable and can be deployed in *IaaS* or *PaaS*. This reduces the deployment footprint, offering flexibility to IT teams. These apps are elastic, with the ability to easily scale in and scale out, which results in optimal utilization of resources. Upgrading, patching, and maintaining cloud-optimized apps is more manageable when compared to *Cloud-Ready* apps. The flip side of *Cloud-Optimized* apps is the learning curve involved in migrating to the cloud. Developers, operators, and system administrators will have to learn the new way of deploying and managing workloads. Line-of-business applications with access to source code are great candidates of being cloud-optimized apps. They can be redesigned and rearchitected to take advantage of the cloud. Cloud-optimized apps can integrate and interoperate with cloud-ready applications.

Cloud-native applications are the most recent breed of applications that are born in the cloud. They take full advantage of the cloud by exploiting capabilities such as elasticity, event-driven resource optimization, and faster release cycles. Apps that are in the early stage of design and architecture are perfect candidates for *Cloud-Native*. *Cloud-Native applications* are designed as *federated microservices*, packaged and deployed as containers, and managed through modern DevOps processes. When your organization needs to large scale to thousands of services in the cloud, microservice application architectures offer major benefits over monolithic modular architectures, including the following:

- Separate *microservices* are better suited to different technology stacks.
- *Federated microservices* enable independent deployment of different parts of the system.
- Isolated deployments add to the resilience and fault-tolerance of the system.
- Scaling characteristics may be different for each microservice.
- APIs are used as connectors.

- Stateless protocols help assure seamless recovery.
- Event-driven programming reduces the overhead of doing needless work and reduces process latency.
- Anti-fragile practices inoculate against unanticipated failure modes.

The target deployment environment for *Cloud-Native applications* includes *Containers-as-a-Service (CaaS)* and *Functions-as-a-Service (FaaS)*. This breed of apps delivers ultimate scalability and availability. Combined with continuous integration and continuous deployment, software is delivered rapidly, which accelerates time to market. It brings closed-loop feedback among developers, testers, operators, and end-users. *Cloud-Native* apps demand new skills of containerization, microservices, and build and release management. However, the ROI is higher when compared to other app models. Greenfield applications that are still in the design phase are ideal for the cloud-native pattern.

!! Key Concept !! Diverse Applications for Cloud Migrations

Large enterprises moving to the cloud will deal with a diverse set of applications. Through hybrid connectivity and integration, *Spaghetti*, *Modular*, *Cloud-Ready*, *Cloud-Optimized*, and *Cloud-Native Applications* will co-exist and interoperate with each other. If you're after the benefits of modularity, make sure you don't trick yourself into a microservices-only mindset. Explore the in-process modularity features or frameworks of your favorite technology stack. You'll get support to enforce modular design, instead of having to just rely on conventions to avoid spaghetti code. Then, make a deliberate choice whether you want to incur the complexity penalty

> of microservices. Sometimes you just must, but often, you can find a better way forward.

In this chapter, it was explained how to engineer applications to be most suitable for DevOps. The next chapter will explain how the pipeline for continuous integration and delivery application changes can be engineered to achieve a well-engineered DevOps environment.

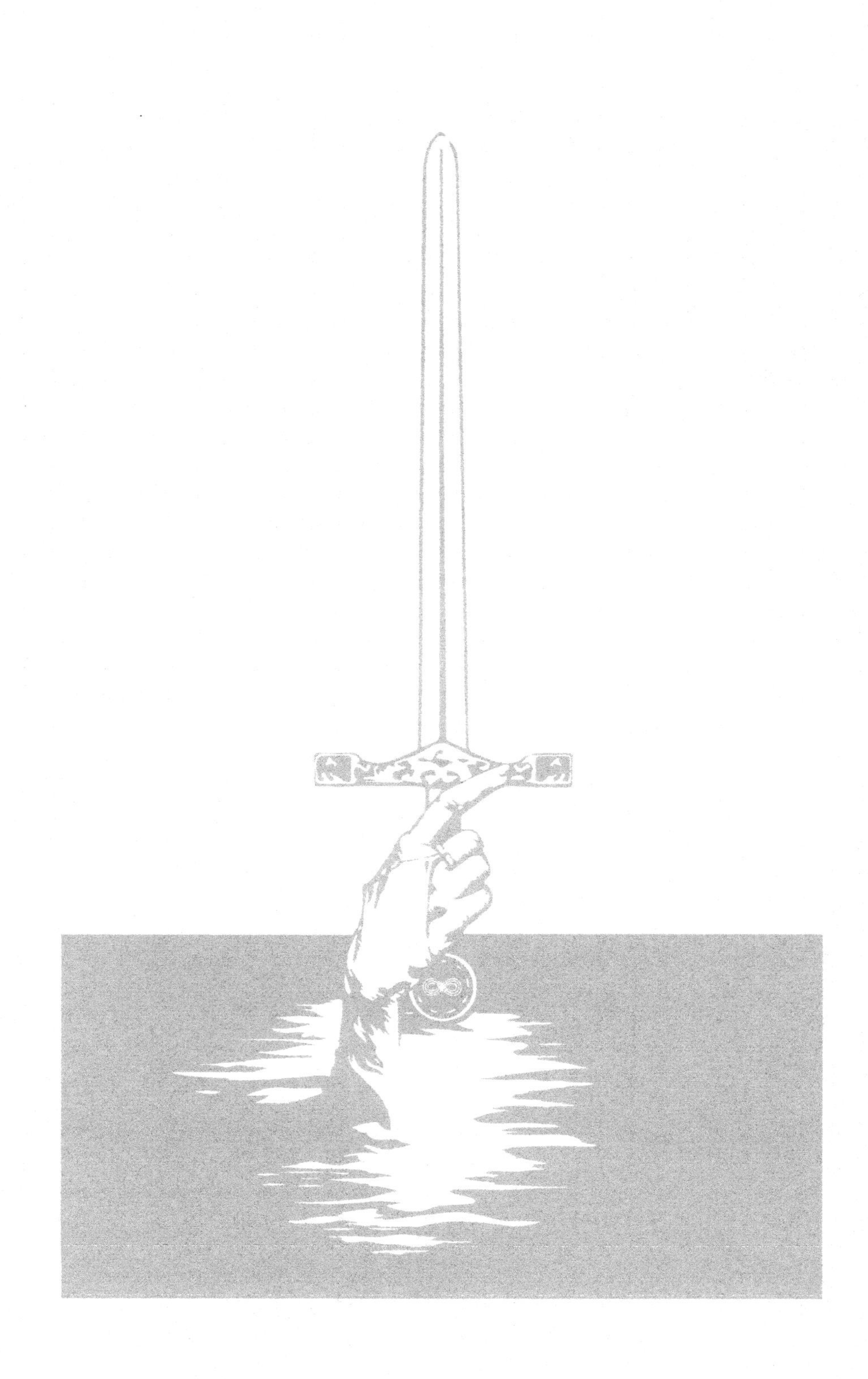

14

CI/CD Pipeline Engineering

As indicated in the last chapter, not only do applications need to be engineered to realize the full potential of a well-engineered DevOps, but the pipeline of continuous integration (CI) and continuous delivery (CD) processes and tools that integrate and deliver changes need to be well engineered as well.

CI/CD pipelines constitute the core of the DevOps engineering blueprint. CI/CD pipelines, also known as "continuous delivery pipelines" in the world of DevOps, refers to the idea that software changes "flow" through a series of stages, from the start of the software planning process until it is delivered to the Operations end. This is analogous to a water purification plant, in which raw water is pumped into the input from a variety of sources, and the combined water inputs flows through valves, couplers, water testers and water treatment processes connected with pipes until the purified water is delivered to consumers. Like the water analogy, processes are meant to be efficient and not impede flow unless a problem is detected.

The specific stages involved in DevOps are not standard. Any organization may have their own definitions and number of stages. **Figure 30—DevOps CI/CD Pipeline** is representative of many continuous delivery pipelines. Most of the time they are represented a series of stages that start with a planning stage and end with deployment to the production operations environment.

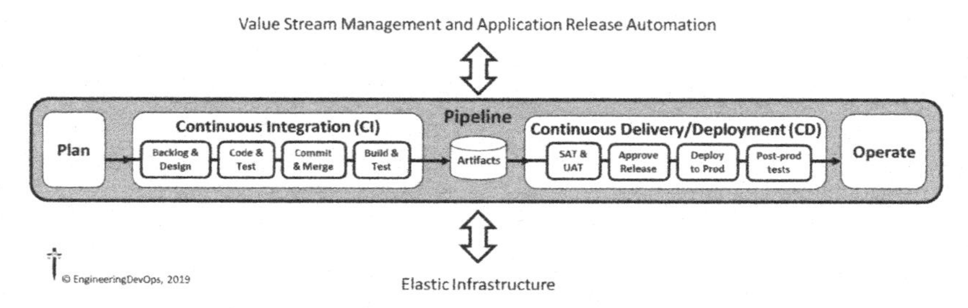

Figure 30—DevOps CI/CD Pipeline

There may be multiple parallel pipelines that interconnect at some points. For example, an enterprise application may consist of a pipeline for application source code and a separate pipeline for database code and data that is used with the application. These two separate pipelines need to intersect at synchronization stages to ensure versions of their artifacts and data are compatible prior to deployment. In enterprises with applications architected with microservice there may be many pipelines—one for each microservice.

Here are some recommended engineering practices that are common to most Continuous Delivery CI/CD pipelines:

- The front-end of the continuous delivery pipeline, called *Continuous Integration (CI)*, is responsible for turning software plans into artifacts that are candidates for delivery.
- The *CI* stage includes scheduling the backlog of user stories and tasks, designing software to meet the assigned stories; coding and unit testing; committing and merging the software changes to a common trunk branch in accordance with a commit protocol; producing a build from trunk; and testing the build with CI tests that typically include at least unit tests, static analysis tests, and some functional tests. Build images that pass CI tests are considered candidates for release and are packaged in a version-managed artifacts repository for easy reference by CD stages.
- Developers test their code changes in their private *"pre-flight"* environment prior to committing them to trunk.

- A software commit by the developer triggers an automated pull request. The changed code is a merged with the latest version of trunk, and CI tests are automatically run. If CI tests pass and all input conditions are met for the integration stage gate, then the merged code is accepted into trunk; otherwise the changes are rejected and flagged for remediation.
- The back-end of the continuous delivery pipeline, called the *Continuous Delivery (CD)*, is responsible for verifying release candidate artifacts, release approval process, preparing for deployment, deployment, and post-deployment testing.
- The CI and CD stages in a well-engineered *Continuous Flow* DevOps environment are automated to such an extent that changes committed by developers that do not require remediation and do not need manual processing to pass all the way to the end of the CD stages.
- While CD usually processes changes to the point of having release candidate artifacts ready for deployment, there may be manual approvals required before deploying to production. This is often the case for platform products that do not favor scheduled deployments and not *Continuous Deployments*. This is the reason CD generally stands for *Continuous Delivery*, not *Continuous Deployment*.
- The CI/CD pipeline is made up of a series of interconnected tools called toolchains.
- The tools used in the *CI/CD* pipeline can be orchestrated and automated (controlled and observed) through APIs by higher-level external tools such as those used for *Application Release Automation (ARA)* and *Value-Stream-Management (VSM)*.
- Tools in each stage of the *CI/CD pipeline* can orchestrate and automate (control and observe) via APIs, the infrastructures that the stage needs to do work.

CI/CD Tools

Each stage in the CI/CD pipeline has specialized tools that perform the tasks assigned to each stage. **Figure 31—CI/CD Tools** shows examples of tools that correspond to the CI/CD pipeline stages. For a more complete list of tools, refer to **Appendix M**.

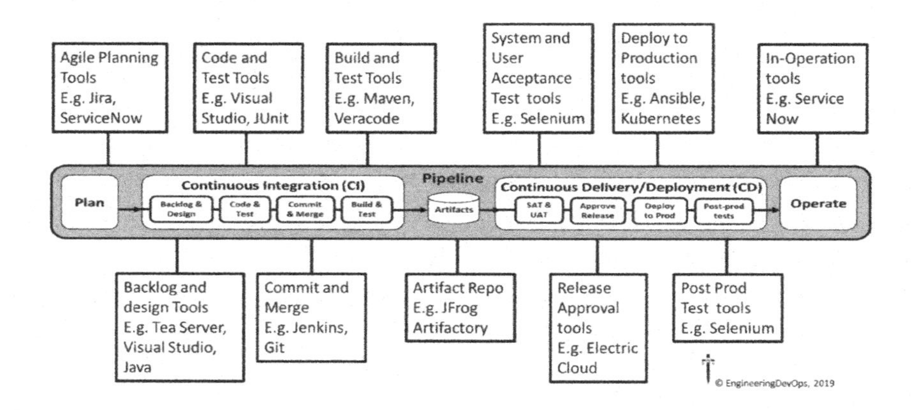

Figure 31—CI/CD Tools

Recommended engineering practices for tools vary depending on the type of tool. For example, the recommended engineering practices for testing tools are different than recommended engineering practices for deployment tools; however, there are some recommended engineering practices common for most types of tools that are used in *CI/CD toolchains*, indicated as follows:

- All capabilities offered by a tool can be controlled and observed though an API.
- Tools can be deployed as software in the cloud. In the case of high-performance tools for testing load and measure latencies that use specialized hardware, a lower-performance configuration is available for deployment to cloud platforms.
- Tools can be deployed in containers easily. *Containerized* tool images are available for popular stacks and container engine variations.
- Tools can be scaled elastically as much as the infrastructure in which they operate.

- Tools are fault-tolerant and have persistence, so they can be recovered without losing state.
- Tools that require a user license have license options that support auto-scaling and do not block scaling requirements when scaling beyond contracted license limits.
- Pre-tested and supported integrations in the form of plugins are available for popular DevOps *CI/CD* tools and *ARA* and *VSM* frameworks such as Jenkins, CloudBees, ElectricFlow, XebiaLabs, Azure DevOps, Harness, and Plutora.
- Metrics to track tool usage and errors are built into the tools and accessible through an API.

CI/CD Toolchains

To realize *Continuous Flow* requires the tools that act on software throughout the *CI/CD pipeline* to be interconnected into a series of tools, like links in a chain, in which the output of tools at earlier stages in the pipeline provide inputs needed by tools in the next and later stages in the chain.

While it is possible for each tool to directly integrate with each other through their APIs, continuous delivery toolchains are usually created by plugging tools into a *CI/CD pipeline*, *ARA*, or *VSM* framework tool such as Jenkins, CloudBees, Azure DevOps, ElectricFlow, XebiaLabs, Harness, and Plutora.

The approach of using a toolchain framework instead of directly coupling individual tools has some essential advantages, detailed as follows:

- The integration into a common *tool framework* reduces the number of pairwise integrations required.
- Off-the-shelf plugins for tools and *tool framework* combinations reduce the cost of integration.
- Common functions such as creating and automating tasks or jobs can be handled by the *tool framework*. This reduces the need for these functions in each tool and provides a common automation paradigm for the entire *toolchain*.

- Tools can be substituted when another tool or tool version is preferred without disrupting the *toolchain* because the *tool framework* can invoke alternatives for specific tools in the chain.
- Scaling the toolchain through a *tool framework* is easier because it can be used to add additional or substitute more powerful instances of tools when needed without disrupting the automation scripts.
- Reliability of the *toolchain* is improved because the *tool framework* can invoke other instances of a tool when needed.
- Monitoring the health of the *toolchain* is improved because the tool framework performs an overseer role for all the tools in the chain.
- Diagnostics and remediation of *toolchain* problems are improved because the *tool framework* can access maintenance capabilities of each tool in the chain through their API.
- Entry and exit *gate criterion* between *stages* in the *toolchain* can more easily implement *policy-as-code* because the decision logic is common across the *tool framework.*
- *Tool frameworks* can be configured in layers to realize the benefits of abstraction such economies of scope and scale. As indicated in the **Figure 19—Application Release Automation (ARA) Engineering Blueprint**, CI *toolchains* and CD *toolchains* can be managed by *ARA tools.* This layered configuration removes the burden of release management from individual CI and CD chains. This is important when and application release requires coordination of multiple CI/CD pipelines. Similarly, *VSM* tools can be layered over *ARA* and *CI/CD tools* to provide overseer capabilities for end-to-end *governance* of multiple *releases* and *toolchains.*

CI/CD with Multitier Applications

The serial, one-way representation of a pipeline is an over-simplification for many organizations and applications. There are often *branches*, loops, and parallel structures in many real *value streams.* For example, some enterprises with applications that use the traditional *three-tier architecture* structure have separate teams working on the Presentation Layer, Business Layer, and Data Access Layer. Each of these layers may have their own pipeline, as shown in the **Figure 32—CI/CD Pipelines for Three-tier App**.

While the *three-tier architecture* is now considered obsolete[RW40] and being replaced by more modern and *Mesh Application Service Architectures*, it does illustrate the point that organizations need to consider how they will manage multiple parallel *Continuous Delivery* pipelines to deliver applications that are partitioned in layers, tiers, or modules that have separate teams and separate workflows. If managing one pipeline is a challenge, you can imagine that management of multiple pipelines that need to be synchronized to make an application *release* is considerably more challenging. The choice of tools and tool *frameworks* need to be approached strategically to ensure the separate pipelines can be coordinated as directed or choreographed federations.

CI/CD for Databases

Databases offer several challenges for CI/CD pipelines.

In the simplest case, when a *database* is used solely by one application, the database code pipeline can be integrated with the application code pipeline to ensure its changes use the same version management system and change controls to ensure the database changes are kept in sync with the application. This approach can also apply to *microservices* in which the data for each *microservice* is bounded within the *microservice*.

If a *database* is used by several applications, then changes to the codified *database* structure must be coordinated with all applications that depend on the *database* structure. Failures in changes introduced in the *database* code need a *database* pipeline that can orchestrate *database roll-backs* and forward deployments in concert with application pipeline versions. In this case, as shown in the **Figure 33—CI/CD Pipeline for Database**, a separate *database* pipeline that is linked to the application pipelines is a good practice.

Database DevOps tools such as Datical, DBMaestro, and Redgate are examples of tools that are designed to support database pipelines. These

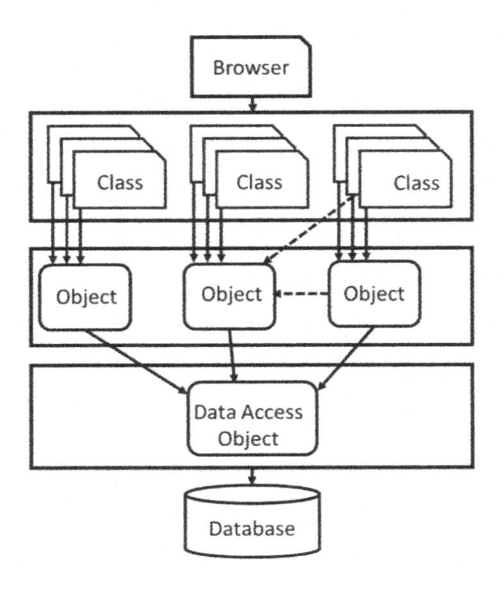

Presentation Layer

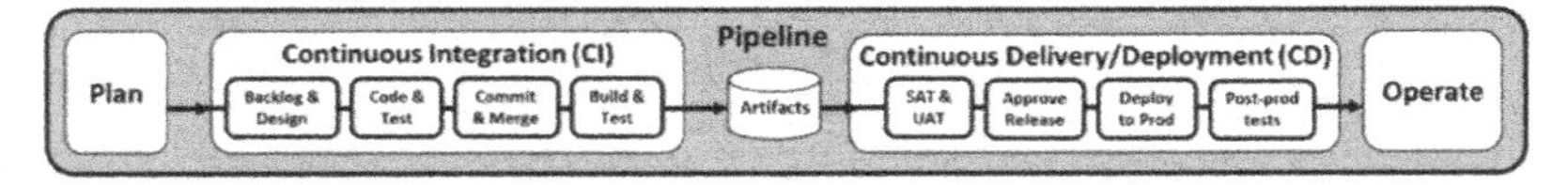

Business Layer

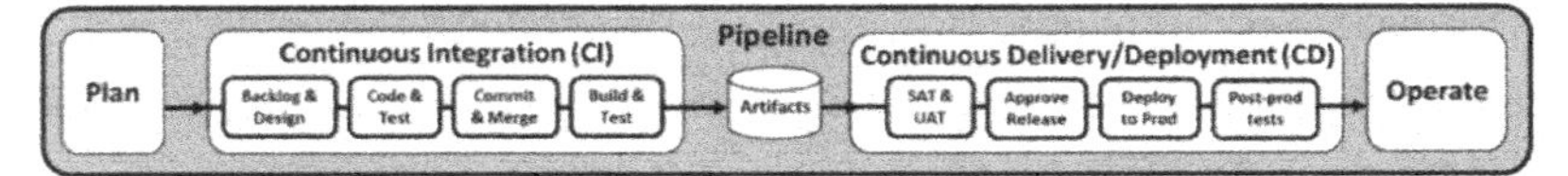

Data Access Layer

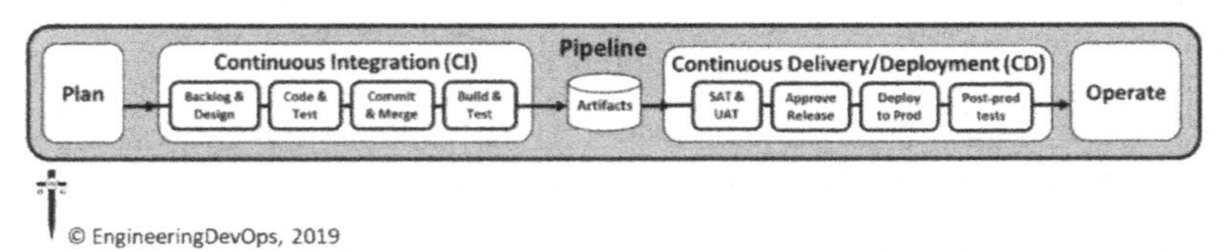

Figure 32—CI/CD Pipelines for Three-tier App

can be integrated with other *CI/CD*, *ARA*, and *VSM tool frameworks* when it is needed to coordinate changes between a database and applications that use the *database*.

Another challenge with *database* pipelines is management of the data in the d*atabase* during the CI and CD stages. Data in production in data needs to be persistent and is ever-changing, even while the underlying *database* code is changing. During the CI process, sample data from production needs to be used to ensure testing is realistic, and new test data must be introduced to verify new *database* capabilities are working.

During the CD stages roll-backs to prior versions of *database* code must be careful to avoid rolling back to versions that are no longer compatible with production data. This is a key reason why *database* changes must follow a disciplined approach that fully tests database changes with applications prior to full deployment. Rolling deployments in which applications and *database* changes are deployed gradually in separate clusters are strongly recommended to minimize the blast radius in case of problems.

CI/CD for Microservices Pipelines

Microservices architectures have proven to be the most mature approach for modern application designs with DevOps. *Microservice* architectures support experimentation with minimal risk provided the infrastructure and practices are mature enough to support the complexity of many parallel *microservices* DevOps pipelines.

Figure 34—CI/CD for Microservices Pipelines shows how successful implementation of *microservices* requires DevOps *CI/CD pipelines* to work as a federation together in a directed or choreographed concert with the microservice architecture of the application and the organizations responsible for them.

The following are some key practices that are critical for operating *CI/ CD* for *microservices* pipelines.

Architecture

- *Microservices* use modular distributed application design practices with strict boundaries controlled and observed solely through APIs communication across a mesh network.

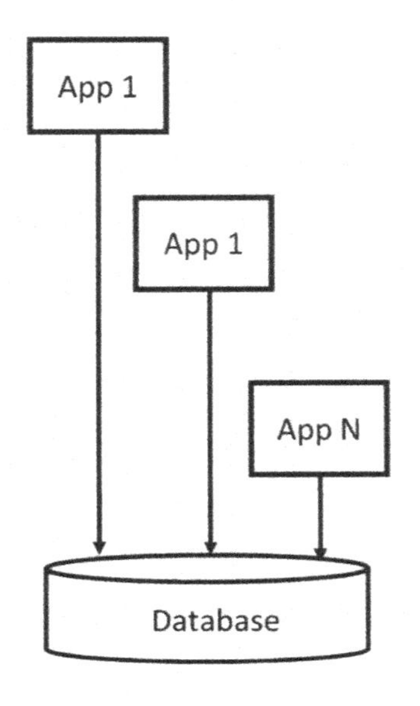

App 1 Pipeline

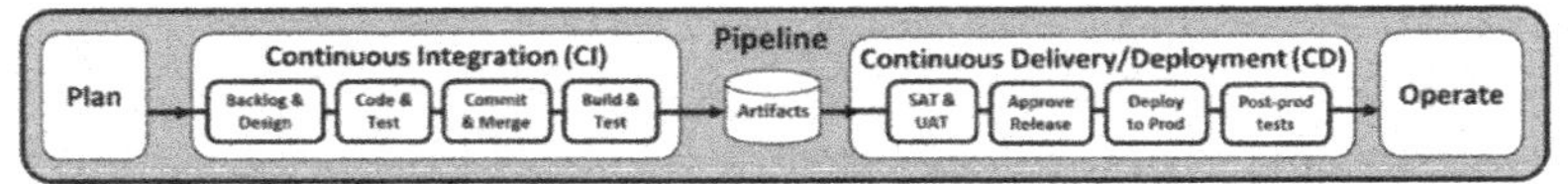

App 2 Pipeline

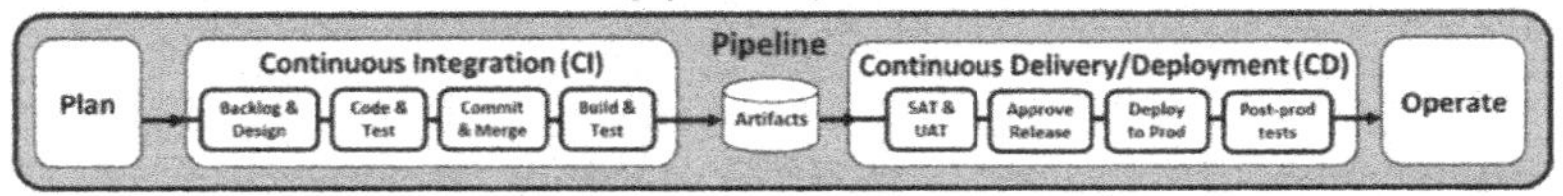

App "N' Pipeline

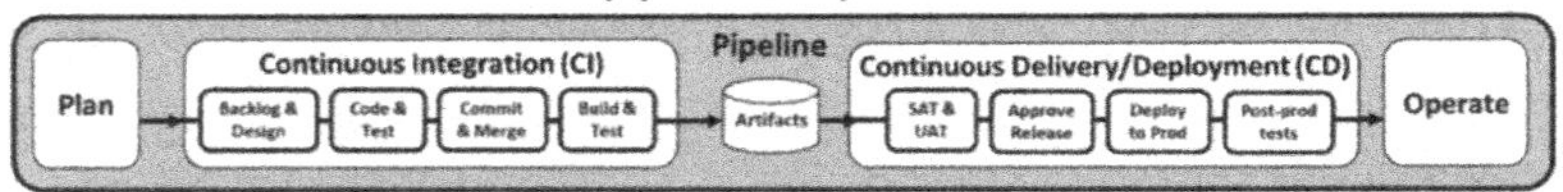

Database Pipeline

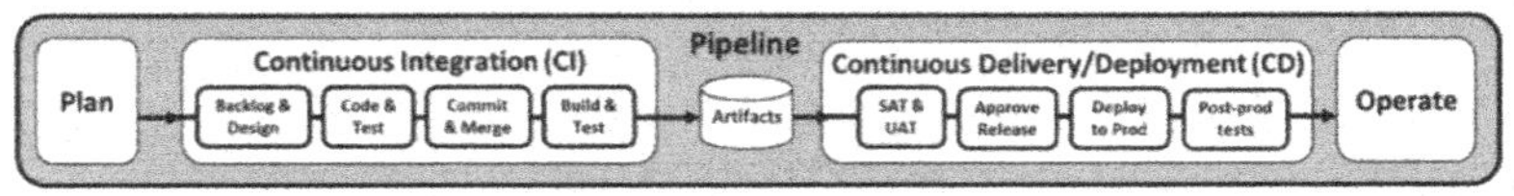

Figure 33—CI/CD Pipeline for Database

- Database scope is contained with the boundary of the microservice.
- Feature Toggles can turn on/off features for testing and deployment.
- Microservices are containerized to support immutable infrastructure practices.

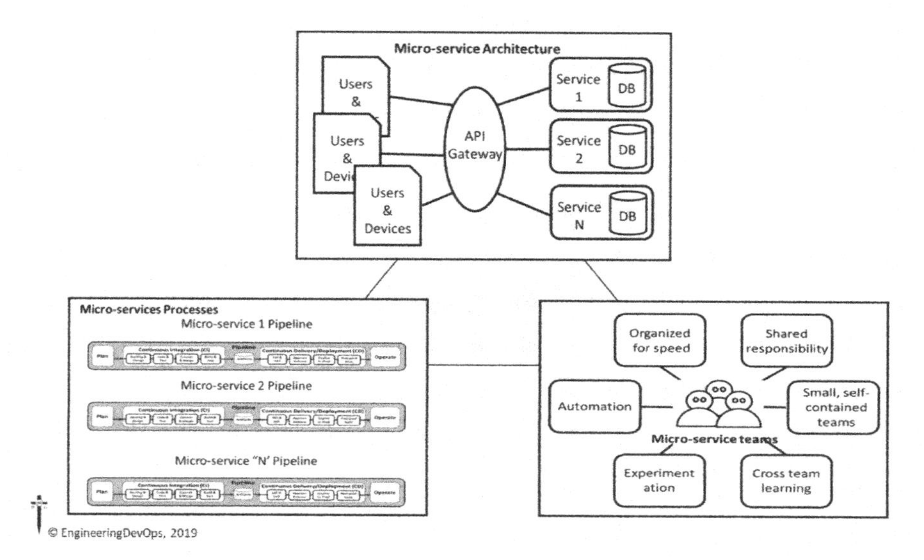

Figure 34—CI/CD for Microservices Pipelines

Process

- Multipipeline *release* automation is used to automatically track and control changes that must be coordinated across multiple *microservices* pipelines.
- *Blue/Green* deployments, *A/B testing*, *Canary* testing, *Continuous Security*, and *Chaos* engineering practices are used to minimize the blast radius and control time to recover to prevent a failed microservice from causing a cascading failure evens in production.
- Each pipeline leverages *elastic infrastructures* with and *orchestration* and automation.
- Multiple *value-stream* orchestration and monitoring supports end-to-end analysis and *governance* of the federated *microservices* for applications.

Organization

- Teams are organized around speed, not efficiency.
- Teams are small and self-contained to match the *microservices* they are responsible for in accordance with *Conway's Law*.
- Cross-team learning within and between teams is proactively pursued.
- Incentives and rewards support experimentation.

CI/CD Pipelines in the Clouds

Considering the global movement of IT infrastructures to the Cloud it is not surprising that all DevOps tools and frameworks are also aiming to run in the Cloud. As indicated in the **Figure 35—CI/CD Pipelines in the Clouds**, all major cloud service providers and CI/CD tool vendors are offering to support DevOps pipelines in the clouds.

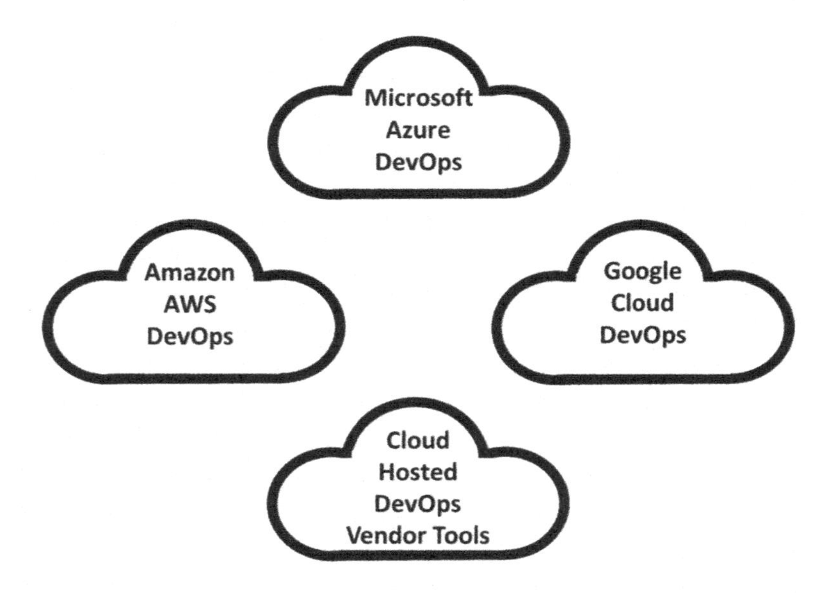

Figure 35—CI/CD Pipelines in the Clouds

There are pros and cons between cloud-provider DevOps pipeline solutions and *Cloud* hosted DevOps vendor *toolchains* made up of separate vendor solutions.

Examples of *cloud* Provider DevOps solutions include the following:

- AWS DevOps includes CodeShip, CodeDeploy, DevOps training and certifications
- Google DevOps includes Google App Engine, Kubernetes, Stack-driver monitor, Stackdriver Debugger, Stackdriver logging, and a Service Catalog that offers a variety of toolchains and solution guides for DevOps on Google use cases.
- Microsoft *Azure DevOps*, an evolution of *Visual Studio Team Server (VSTS)*, includes *Azure Boards*, *Azure Pipelines*, *Azure Repos*, *Azure Testing*, *Azure Artifacts* and a marketplace for third-party tool plugins.

Pros of *Cloud* Provider DevOps pipeline solutions include the following:

- Tools tend to be designed for cloud deployments, and many are cloud-native.
- Pay-for-use billing allows scaling the pipeline while controlling costs.
- Global *Cloud* networks offer DevOps solution located in each cloud region.
- *Cloud* provider pipeline tools provide performance guarantees (SLAs).
- Service and support for pipelines are provided by the cloud service provider.
- Integration with other services are offered by the Cloud service provider.

Cons of *Cloud* Provider DevOps pipeline solutions include the following:

- Lock-in to *Cloud* vendor solutions. Each *Cloud* provider's solution favor their own Cloud services.
- *Hybrid-Cloud* and *Multi-Cloud* configurations support is not a priority.

Examples of *Cloud* Hosted DevOps Vendor Solutions include the following:

- CodeShip (acquired by CloudBees in 2018) is a hosted CI platform available as SaaS.
- Circle CI is a *Cloud*-hosted CI solution.
- CloudBees is derived from Jenkins and can be hosted on premises or in the *Cloud*.
- ElectricFlow (recently acquired by CloudBees) is an ARA solution that can be installed on premises or in a *Cloud*.
- Harness automated CD pipelines are available as SaaS or on-premises.
- Jenkins is an open-source CI/CD solution that can be installed on premises or in the *Cloud*.
- Shippable (now part of JFrog) is a pipeline tool designed for containers available for on premises or in the *Cloud*.
- Travis CI is a *Cloud*-hosted CI solution alternative to *Jenkins*.

Benefits of *Cloud*-Hosted DevOps *toolchain* solutions include the following:

- Users are free to choose tools, *tool frameworks*, and craft *toolchains* from any number of vendors or open sources.
- Feature evolution for specific vendor tools is faster.
- Support for *Hybrid-Cloud* and *Multi-Cloud* scenarios are provided.

Disadvantages of *Cloud*-Hosted DevOps *toolchains* solutions include the following:

- Users are locked into specific tool vendor solutions.
- Users need to create and maintain their own *toolchain*.
- Users need to manage deployments and regional support for the *toolchains*.
- Scaling of some tools may not be sufficient for some tools.

Five Levels of CI/CD Pipeline Maturity

The maturity of *CI/CD pipelines*, as shown in **Figure 36—CI/CD Pipeline Maturity Levels**, is indicted by the sophistication of the toolchains their integrated workflows.

At the lowest level, **Chaos Reigns**, the *value stream* does not have a fully connected pipeline or *toolchain* for CI or CD. People use the tools manually. The manual interactions cause bottlenecks and are a source for errors, low repeatability, and low reliability.

At the second level, **Continuous Integration**, the *CI pipeline* stages are supported by an operational *toolchain*. Developers can commit their changes and the system will take care of integration, builds, testing, producing release artifacts, and flagging failures for remediation.

At the third level, ***Continuous Flow***, otherwise known as *The First Way* of DevOps, both the CI and CD pipeline stages are supported by operational *toolchains*. *Release* candidates can be prepared and fully qualified for deployment with minimal manual effort.

At the fourth level, ***Continuous Feedback***, otherwise known as *The Second Way* of DevOps, the end-to-end *CI/CD pipeline* is instrumented with metrics that make visible performance of the pipeline, the applications, and the infrastructure.

At the fifth level, ***Continuous Improvement***, otherwise know at *The Third Way* of DevOps, the *CI/CD pipeline* is reliable, fast, and uses advanced DevOps practices for advanced *continuous testing*, *continuous security*, and advanced *continuous deployment*. The robust, sophisticated pipelines provide confidence for proactive experimentation with new pipeline tools, integrations, and workflows while managing risk.

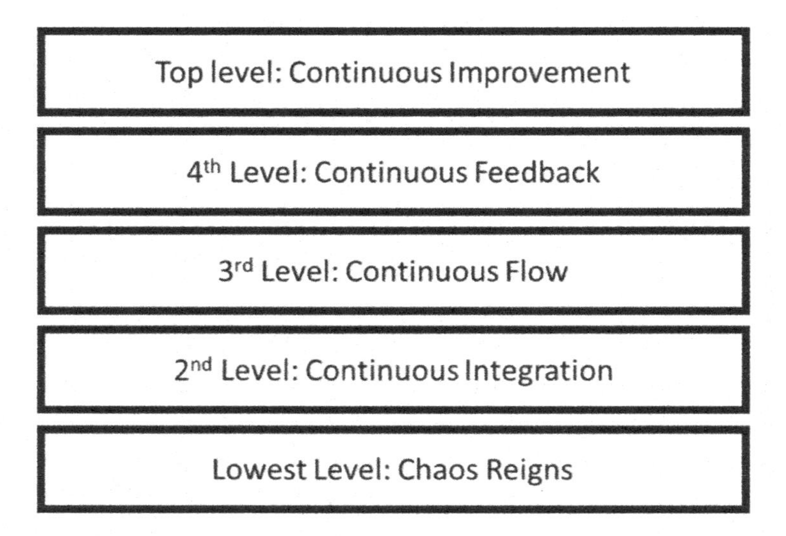

Figure 36—CI/CD Pipeline Maturity Levels

This chapter provided a blueprint and engineering practices for CI/CD pipelines. The next chapter provides blueprints and engineering practices for infrastructures that applications and CI/CD pipelines can operate in.

15

DevOps Elastic Infrastructures

As indicated in the last chapter, DevOps needs infrastructure to run application and pipelines. From a system engineering point of view, this seems obvious. Ironically, customers of DevOps would prefer not to have to think about infrastructures or CI/CD pipelines. They have enough to do developing and operating their applications. The following anecdote is an interesting case in point.

While grabbing a coffee in the break room at one of my client's office buildings, I overheard a software engineer talking to her teammate. She said, "I am a modern software developer. I don't need a computer server. I just create and test my code from my laptop. Somehow infrastructure in the background acts as an invisible service that takes care of my needs for computations, storage, and transfers of information safely. I don't know how it does it. I don't care."

Wow. Things have changed from "the good old days" when, as a summer student working for Nortel in 1977, I had to create the digital circuits, nanocode, firmware, and high-level source code to build a small system for testing trunk interfaces on a packet switch. I remember having to care a lot about the count of opcodes to make sure my programs fit and performed in the sixteen kilobytes of erasable *Read-Only Memory* (eROM) that was the deployment environment for my code. In those days, clouds were bad things because they foretold of nasty rain or snow. If someone back then said clouds had computers in them, people would have worried about hardware falling from the sky. But things changed very fast, even back then. Ten years later I had been promoted several times and had become Senior Director of Global Test Technologies at Bell-Northern Research (BNR). In 1989, I published a paper called "TEAM: A Desktop Window to the Captive Office,"[RR8] which describes a system my team at BNR had developed. As the title implies, TEAM allowed thousands of engineers across the global company to run tests on any of the systems infrastructures deployed in thirteen private (captive) labs across the global company by simply clicking some buttons on their desktop window. (Yes, we did have windows back then too—X-windows for Unix, not Microsoft Windows). The engineers did not have to care or know how the infrastructure did that.

BNR was an advanced engineering environment for its day and had capabilities not available to most engineers working outside such advanced working environments. Today, thirty years later, software engineers everywhere have access to a host of infrastructure services and barely need to think about it (or Opcodes!) while they code and test their applications.

To engineer DevOps, a knowledge of infrastructures, their characteristics, and recommended engineering practices for using and managing them are critical and are outlined in this chapter.

Ephemeral Elastic Infrastructures

DevOps works best when the infrastructure is engineered for developers as a flexible service. If a small, low-risk software change is committed to the trunk, the CI stage needs very little infrastructure (compute, storage, and network) to do the work tasks of building, testing, and packaging the change. When a large or risky change is committed, the CI stage needs much more infrastructure. In either case, when the work tasks are complete, the infrastructure is no longer needed for the completed task.

The same level of work demand variance exists throughout all stages of the pipeline. Bear in mind that for any one application there may be many developers committing code frequently, but asynchronously. These different code commits may be various sizes and risk levels, so the infrastructure needed to build and test them must to be highly flexible. Real infrastructures consist of real computers, storage, and network devices that do not magically appear and disappear depending on work demands. But with modern infrastructures suited for DevOps, those real systems exist in distributed, highly configurable, and secured resource pools that are dynamically configured on demand. From a developer's point of view, these resources appear to be ephemeral and elastic.

Idempotency and Immutable Infrastructure

I have a nightmare in which I drive on the same road, but it leads me to a different destination every time I use it. Same car, same driver. What happened to the road? This can happen to computer systems infrastructures. When the same program is run on computing infrastructure, you can get a different result if the computer system configuration (hardware settings, operating system environment variables, versions of code in the software stack, etc.) has changed between executions. This can happen when one server is patched but not others, or if the patch process did not complete correctly. Over the course of multiple patch updates the configurations drift away from what is intended. There are two words used to describe these characteristics that you need to know to understand elastic infrastructure—idempotent and immutable.

Idempotent *Configuration Management Systems* are designed to prevent infrastructure configuration drift. They proport to guarantee *idempotency*[RW35]—the system configurations for infrastructures remain consistent so that each time they are used you get the same result. *Configuration Management Systems* such as *Puppet* and *Chef* were designed around the concept of idempotency. They scan infrastructure components to validate their configurations match what they are programmed to expect and update configurations that are not compliant. These systems are a boon for infrastructure managers who need to ensure the infrastructure does not drift between uses. However, there are several problems with this approach. The configuration update process can take a long time if there is a large stack of code to be imaged. And every time a configuration image is built, there is a risk that unintended changes can creep in. Unintended changes can cause functional, performance or security problems. If this occurs, the idempotent tools ensure bad configurations are on every server!

Immutable infrastructures refer to an approach to managing services and software deployments on IT resources wherein components are replaced rather than changed. An application or services is effectively redeployed each time any change occurs.[RB17] *Immutable* infrastructure benefits include lower IT complexity and failures, improved security, and easier troubleshooting than on *mutable* infrastructure. It eliminates server patching and configuration changes, because each update to the service or application workload initiates a new, tested, and up-to-date instance. There is no need to track changes. If the new instance does not meet expectations, it is simple to roll back to the prior known good instance. Since you're not working with individual components within the environment, there are far fewer chances for unpredictable behaviors or unintended consequences of code changes.[RW36]

Bare Metal, Virtual Machines, Containers, and Serverless

The workhorse of IT is the computer server that software application stacks run on. The server consists of operating system, computing, memory, storage and network access capabilities often referred to as a computer machine or just "*machine.*"

A **bare metal machine** is a dedicated server using dedicated hardware. Data canters have many *bare metal servers* that are racked and stacked in *clusters*, all interconnected through *switches* and *routers*. Human and automated users of a data center access the machines through access servers, high security *firewalls*, and *load balancers*.

The **virtual machine (VM)**, such as those supported by VMWare™, introduced an operating system simulation layer between the *bare metal server's* operating system and the application so one *bare metal server* can support more than one application stack with a variety of operating systems. This provides a layer of abstraction that allows the servers of a *data center* to be software-configured and repurposed on demand. In this way, a virtual machine can be scaled horizontally by configuring multiple parallel machines or vertically by configuring machines to allocate more power to a *virtual machine*. One of the problems of virtual machines is the *virtual operating system* simulation layer is quite "thick," and the time required to load and configure each VM typically takes some minutes. In a DevOps environment, changes occur frequently. This load and configuration time is important because failures that occur during the load and configuration can further delay the instantiation of the *VM* and application. The *idempotent* characteristic of modern *Configuration Management* tools such as Puppet, Chef, Ansible, and SaltStack do help to reduce the chance for errors, but when configuration errors are detected, delays to reload and reconfigure the stacks take time. These delays constitute significant wait time for each DevOps stage, and the accumulated time for a series of stages needed for a DevOps pipeline can be a significant bottleneck. *VMs* are used in production IT environments because of the flexibility to support different configurations. However, the time to reload and configure *VMs* can be a bottleneck during release reconfigurations and can delay mean-time-to-repair when failures occur.

Containers, such as those supported by Docker™, have a very lightweight operating system simulation layer (the Docker™ ecosystem) that is tailored specifically to each application stack. As shown in the **Figure 37—Containers**, container systems such as Docker guarantee isolation between application stacks run over the same Docker layer. The smaller "footprint" of containerized application stacks compared to *VMs* allow more application stacks to run on one bare metal machine or virtual machines, and they can be instantiated in seconds rather than minutes. These desirable characteristics make it easy to justify a migration from running application stack on *bare metal machines* or *VMs*. Instead of time-consuming loading and configuring an application stack and OS layer, for each machine instantiation a complete container image is loaded, and a small number of configuration variables are instantiated in seconds. This capability to quickly create and release *containerized applications* allows for the infrastructure to be *immutable.*

In thirty minutes, someone can go from having nothing on their machine to a full dev environment and ability to commit and deploy. *Containers* are the fastest means to launch short-lived, purpose-built testing sandboxes as part of a *Continuous Integration* process. As shown in **Figure 38—Containers Workflow**, *containers* are the result of a build pipeline (artifacts-to-workload). Test tools and test artifacts should be kept in separate *test containers. Containers* include minimal runtime requirements of the application. An application and its dependencies bundled into a *container* is independent of the host version of Linux kernel, platform distribution, or deployment model.

Benefits of containers include the following:

- The small size of a container allows quick deployment.
- Containers are portable across machines.
- Containers are easy to track and easy to compare versions.
- More containers than virtual machines can run simultaneously on a host machine.

The challenge with *containers* is to get your application stack working with Docker. Making *containers* for a specific application stack is a project easily justified by the ROI of reduced infrastructure costs and the benefits provided by *immutable* infrastructures such as fast *Mean-Time-To-Restore-Service (MTTRS)* during failure events.

Containerized infrastructure environments sit between the host server (whether it's virtual or *bare-metal*) and the application. This offers advantages compared to *legacy* or traditional infrastructure. *Containerized* applications start faster because you don't have to boot an entire server. *Containerized* application deployments are "denser" because *containers* don't require you to virtualize a complete operating system. *Containerized*

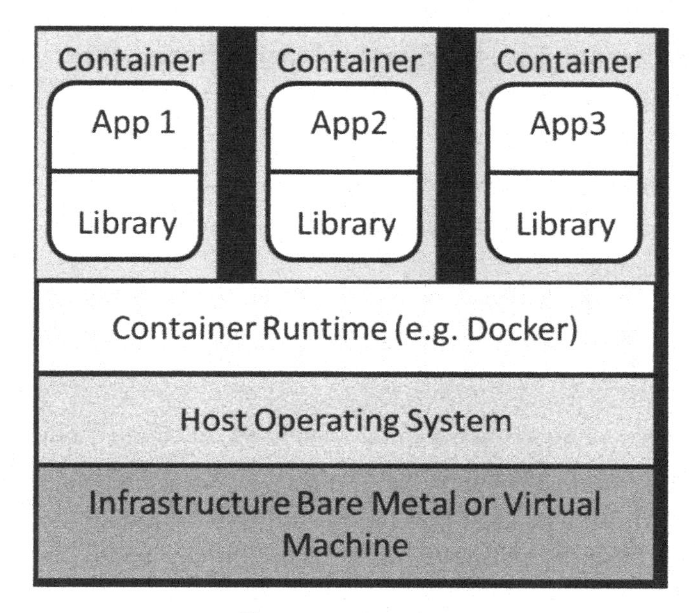

Figure 37—Containers

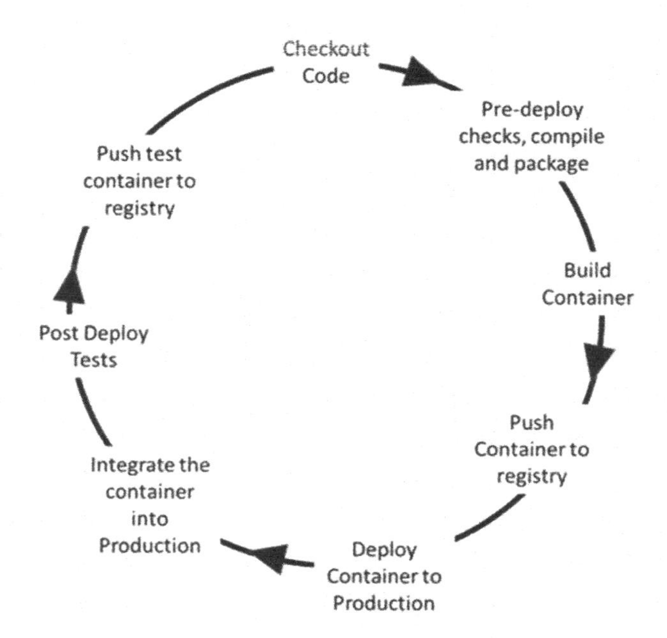

Figure 38—Containers Workflow

applications are more scalable because of the ease of spinning up new *containers*. Here are some recommended engineering practices for *containerized* infrastructure:

- *Containers* are decoupled from infrastructure.
- *Container deployments* declare resources needed (storage, compute, memory, network).
- Place specialized hardware *containers* in their own cluster.
- Use smaller clusters to reduce complexity between teams.

Serverless computing is a *cloud-computing* execution model in which the cloud provider runs the server and dynamically manages the allocation of machine resources. Pricing is based on the actual amount of resources consumed by an application, rather than on pre-purchased units of capacity.[RW37] *Serverless computing* can simplify the process of deploying code into production. Scaling, capacity planning, and maintenance operations may be hidden from the developer or operator. *Serverless* code can be used in conjunction with code deployed in traditional styles, such as *microservices*. Alternatively, applications can be written to be purely *serverless* and use no provisioned servers at all.

Infrastructure as Code (IAC)

Infrastructure as Code (IaC), as illustrated in **Figure 39—Infrastructure-a-Code (IaC)**, is the management of infrastructure (*networks*, *virtual machines*, *load balancers*, and connection topology) using descriptive [codified] models, managed with versioning same as source code. IaC provides the following benefits:

- Reproduces configurations automatically
- Generates the same environment every time it is applied
- Used in conjunction with DevOps/*Continuous Delivery*

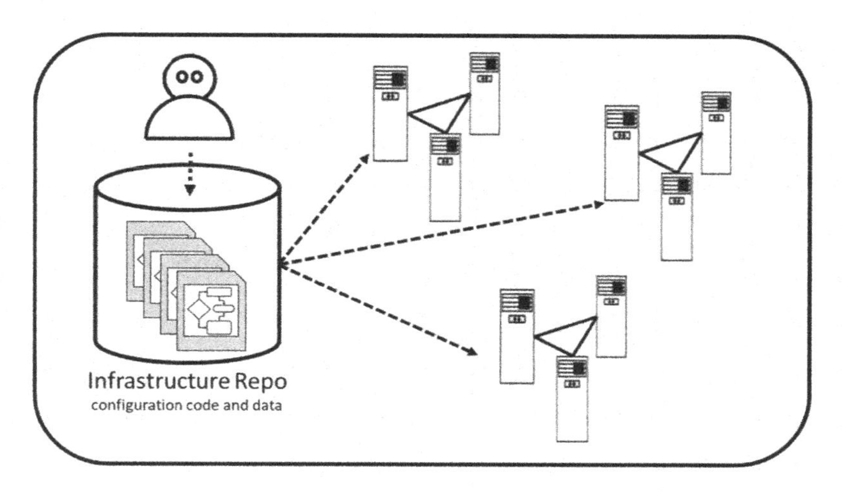

Figure 39—Infrastructure-a-Code (IaC)

DevOps *IAC* automation use cases include the following:

- *Pre-Flight* test environment
- Staging automation
- Deployment automation
- Safety: *Green/Blue* deployments
- Quality: *A/B testing* methodologies
- Restore and recovery
- Cost control
- Utilization
- *Governance*

IaC works by leveraging the following capabilities:

- *Idempotency*—automatically configure infrastructures
- *Immutability*—discard and recreate fresh infrastructures
- *Ephemeral*—virtual resources that are created and released on-demand
- *Orchestration*—create and release infrastructure resource
- *Automation*—execute tasks

- Ease of use and admin—unified practices and tools
- Speed—deliver stable environments rapidly
- Scale—vertically and horizontally on demand
- Repeatability—resolve environment drift in the release pipeline
- Maintenance—version-manage infrastructure code and data

Infrastructure-as-Code Tools

Tools, *orchestration*, and *automation* are needed to elastically scale and de-scale infrastructure depending on traffic and capacity variations while maintaining control of costs. **Figure 40—Infrastructure-as-Code Tools** shows some popular DevOps pipeline automation tools, infrastructure configuration, and *orchestration* tools and *ephemeral* infrastructure tools and services.

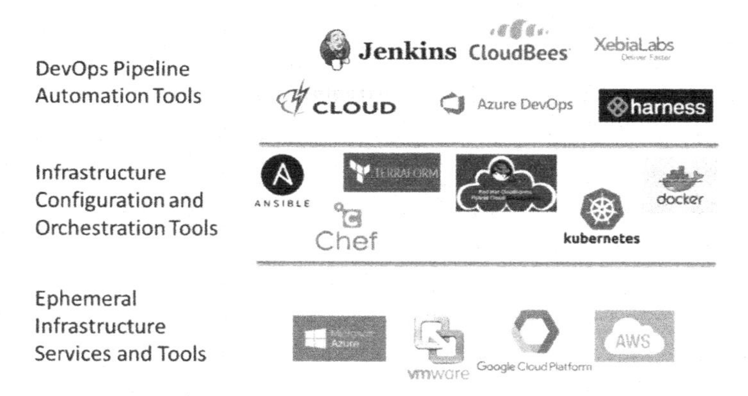

Figure 40—Infrastructure-as-Code Tools

The following recommended engineering practices associated with *Infrastructure-as-Code* are presented according to the Three Dimensions of DevOps—People, Process, and Technology.

People engineering practices for IAC are as follows:

- The infrastructure is created by an expert team and reviewed by a coalition of stakeholders including representation from Dev, Ops and QA roles.

- The user performance of build and test processes experienced by different teams are consistent with SLAs determined for each team. SLAs and monitoring tools measure the user performance experience for all teams.
- There is *Role-Based Access Control (RBAC)* to the code and IaC tooling to follow the *Least-Privilege* model.
- "Codify everything" is a concept that serves to develop a culture where all infrastructure and configuration is defined.
- Identify champions, often outside of the direct purview of the DevOps team; align executive sponsors.

Process engineering practices for IAC are as follows:

- Infrastructure changes are implemented following a phased implementation process to ensure the changes do not disturb the current DevOps operation. Examples of implementation phases in include Dev, Test, Stage, and Prod.
- *Key Performance Indicators (KPIs)* to monitor the performance of the DevOps infrastructure are defined and in use.
- Logs and proactive system alerts are in place for most infrastructure failures and are organized in a manner to quickly identify the highest-priority problems.
- *Blue/Green* deployment methods are used to verify releases in a controlled deployment environment (e.g., Blue) before activating the environment to live (e.g., Green).
- *"A/B testing"* methods to trial different versions of code features with customers in separate live environments are supported.
- Infrastructure supports *"canary testing"* methods to trial new code versions on selected live environments.
- Metrics and thresholds for the infrastructure components are automatically gathered, calculated, and made visible to anyone on the team that subscribes to them. Example metrics are availability (up time) of computing resources for CI, CT, CD processes; time to complete builds; time to complete tests; number of commits

that fail; and number of changes that need to be reverted due to serious failures.
- Infrastructure failure modes are frequently tested for resiliency. *IaC* serves to implement not only infrastructure but also the *governance* and *security* decided on in architecture, and it is continuously evolved, tested, and integrated.

Technology engineering practices for IAC are as follows:

- Infrastructure is available on demand, as code, to support testing software changes by developers in a private environment prior to committing the software changes to the integration/trunk branch.
- A *Software Version Management (SVM)* system (e.g., Git) is used to manage all versions of tools and infrastructure configurations for build and test processes.
- Infrastructure for builds and tests are available sufficiently that they never block a build.
- Infrastructure resources for build and test processes can be scaled up and down automatically under the control of an API.
- *System Configuration Management* and system inventory data is stored and maintained in a *CMDB*.
- Infrastructure changes are managed and automated using configuration management tools that assure idempotency.
- Automated tools are used to support immutable infrastructure deployments.
- Build, test system and deployment system fault monitoring, fault detection, system and data monitoring, and recovery mechanisms are automated.
- Infrastructure disaster recovery procedures are automated and verified.
- *Secrets management* is provided as a service to avoid storing credentials and credential systems in code.

NetDevOps

NetDevOps extends capabilities of DevOps into the network. As networks begin to embrace more software-defined networks (SDN) and virtualized network functions (VNFs) for delivery of networks as services (NaaS), the ability to enable and manage network changes must become a core competency. This core competency requires deep understanding of DevOps engineering practices that enable automated deployments of network elements and network topologies from development, pre-deployment testing, and into production. NetDevOps methods provide the required best practices to realize network elements as software, images and configuration files over ephemeral cloud infrastructures or real physical hardware, or a combination of cloud and physical resources.

Enabling automated network services in this new networking paradigm still requires the use of legacy physical systems to cooperate with next-generation virtual network elements, many of which will operate on whitebox hardware or somewhere in a data center on the network. The hybrid networking requirement makes what is an already very complex, heterogeneous environment to test even more complex. In these cases, the importance of infrastructure as code and cloudification of hybrid test environments such as lab-as-a-service and test-as-a-service platforms become very effective in delivering NetDevOps pipelines and productive continuous test routines. Appendix Q includes a NetDevOps Blueprint.

Ad-Hoc Infrastructures

Ad-hoc is the least mature type of infrastructure. Machines exist as *bare metal* or *virtual servers* in a computer facility. Communication between machines is usually very limited and restricted to direct communication between specific subsets of machines. There is no centrally managed infrastructure communication structure that allows all machines to in the computer facility to communicate with each other. Machines are usually managed manually as individual *bare metal* or *virtual servers* without a central *Configuration Management System*. *Security* is less sophisticated compared to other infrastructures and usually depends on physical access security to the computer center. Ad-hoc infrastructures offer low cost for infrastructure management tools and expertise, but this is traded off for the flexibility and efficiencies afforded by more mature infrastructures. Ad-hoc infrastructures are usually not suitable for DevOps.

Private Data Centers

Private data centers are dedicated computing facilities for exclusive use by an organization. *Private data centers* have been the mainstay of enterprise computing for many years. The facilities may be co-located with the consumers of the resources or located elsewhere and consumed remotely. They are highly structured, centrally managed networks of computing and storage capabilities that may consist of a combination of *bare metal*, *virtual*, or *containerized machines*. A *private data center may* offer services to its users including *auto-scaling* within the limitations of the installation. *Private data centers* can offer very high security because they are private. The capital and operating costs of *private data centers* are fixed and borne by the organization that owns it. Management, capacity planning, and technology evolution is the responsibility of the same organization that consume its capabilities. *Private data centers* are suitable for DevOps; however, Cloud Services are preferred.

Cloud Services (IaaS, PaaS, SaaS, FaaS)

A ***Cloud Service*** is any **service** made available to users on demand via the Internet from a cloud computing provider's server as opposed to being provided from a company's own on-premises servers.

Cloud services include: ***Infrastructure-as-a-Service (IaaS)***, ***Software-as-a-Service (SaaS)***, ***Platform-as-a-Service (PaaS)***, and ***Function-as-a-Service (FaaS)***.[RB18]

Infrastructure-as-a-Service (IaaS)[RW34] online services provide high-level APIs used to dereference various low-level details of underlying network infrastructure like physical computing resources, location, data partitioning, scaling, security, backup, etc. A *hypervisor*, such as Xen, Oracle VirtualBox, Oracle VM, KVM, VMware ESX/ESXi, Hyper-V, or LXD, runs the virtual machines as guests. Pools of *hypervisors* within the cloud operational system can support large numbers of virtual machines and the ability to scale services up and down according to customers' varying requirements.

Software-as-a-Service (SaaS) is a capability in which a service provider makes software applications available over the internet.

Platform-as-a-Service (PaaS)[RW33] is a category of cloud computing that provides tools and application stacks that enable applications to be developed, run, and managed without the complexity of building and maintaining the infrastructure typically associated with developing and launching an app.

Function-as-a-Service (FaaS),[RW38, RW39] also known as ***serverless architecture***, provides developers with the opportunity to create applications without thinking about the infrastructure behind. These services take the *Platform-as-a-Service (PaaS)* concept to a new level where users don't have to manage servers. The machines needed for a service are created and released elastically according to event-driven work demands, without the users having to manage servers. FaaS offers characteristics that are very well suited for high performance DevOps, including the following:

- Reduced costs due to much lower hold times of infrastructure resources
- Reduced management overheads
- Reduced development and deployment costs
- Faster time to market for innovations
- Auto-scaling
- Fault-tolerant
- Self-healing

Example recommended engineering practices for cloud services are listed in the following subsections for the following categories:

- Cost Management
- System Performance
- High Availability
- Change Management
- Security and Compliance

Cloud Cost Management Recommended Engineering Practices

- User accounts management and alerting prevents users from over-provisioning.
- Managed services are used to reduce administrative and operational costs.
- Benchmarking assessments ensure that resources are optimized to match workloads.
- A pricing model is selected to match workloads to the least-cost solutions offered by the cloud provider, such as a mix of on-demand instances, reserved instances, and interruptible instances.
- *Serverless* functions are used for applications that execute only on set schedules or when triggered by an event.
- Policies and procedures to monitor, control, and appropriately assign your costs leverage cloud provider-provided tools for visibility.

Cloud System Performance Recommended Engineering Practices

- *Network connectivity*, *address management*, and *name resolution* reduce risk of contention.
- *Load balancers* and *auto-scalers* auto-flex resources to match variable workload demands reliably.
- *Cloud Solutions* improve application performance by engineering enhanced network performance, *VM* size, data systems performance, and routing table optimizations.

- Connectivity options between *private data centers* and *cloud providers* use solutions such as *VPN*, *Peering*, *Direct Connections*, and *Express Route* to meet traffic transfer needs.
- Multilabel Protocol Switching (MLPS) minimizes the number of connections needed between multiple data centers that are to be connected to the cloud.
- Compute solutions performance engineering considers requirements of application design, usage patterns, and configuration settings.
- Migrations of application from a data center to the cloud using some application refactoring or replatforming will yield improvements but may take more investments.
- Storage solutions performance engineering considers access methods (block, file, or object), patterns of access (random or sequential), throughput required, frequency of access (online, offline, archival), frequency of update (*WORM*, dynamic), and availability and durability constraints.
- Database solutions performance engineering considers requirements for availability, consistency, partition tolerance, latency, durability, scalability, and query capability.
- Network solutions performance engineering considers latency, throughput requirements, and physical constraints.
- New capabilities are introduced as new technologies and approaches become available that could improve the performance of the architecture.
- Monitor performance of systems to identify degradation and remediate internal or external factors (such as the OS or application load).
- *Cloud solutions* performance engineering consider trade-offs (such as consistency, durability, space versus time, latency, etc.) to deliver higher performance by sacrificing some other constraint.
- Solutions are engineered to balance spend and performance to prevent both over-utilization and over-provisioning.

- Data-transfer charges are monitored to ensure architectural decisions alleviate unnecessary data transfer costs.

High-Availability Cloud Services Recommended Engineering Practices

- Redundant services available in multiple regions support high-availability disaster recovery.
- Logs, metrics, and thresholds provide visible health checks and self-healing of cloud services, resources, and applications.
- *Mean-Time-To-Restore-Service (MTTRS)* performance metrics are defined, measured, and acted upon for infrastructure and data.
- Applications are distributed across physical locations (zones and regions) to support availability goals.
- Routinely practice incident response scenarios with tools and workflows that ensure investigation and recovery practices are kept current and efficient.
- Support expected from cloud providers is contracted, such pay-per-incident to unlimited incident support.
- *SLAs* with business units are aligned with *SLAs* available from cloud service providers.
- Unplanned operational events are supported by pre-defined playbooks, escalation protocols, *automated* responses for alerting, mitigation, remediation, roll-back, and recovery.

Cloud Change Management Recommended Engineering Practices

- Change controls tools manage cloud service configuration changes.
- Implement change control and resource management so that you can identify necessary process changes or enhancements where appropriate.
- Changes to configuration of stacks and configuration parameters for resources are tracked and automated.

- Changes to cloud capabilities are implemented using automation, small frequent changes, regular quality assurance testing, and defined mechanisms to track, audit, roll back, and review changes.

Cloud *Security* and *Compliance* Recommended Engineering Practices

- *Cloud account credentials* are managed with identity and access management.
- *Role-Based Access Controls (RAC)* are used for cloud user accounts, user-groups, and automated consumers of cloud services.
- *Firewalls* are used to *secure* boundaries of subnets, *VMs*, and operating systems.
- *Data* is protected using access controls, regional boundaries, and encryptions according to sensitivity and class of data.
- Protect data at rest using encryption and segregation in buckets and volumes.
- Cloud access authentication integration with other authentication systems reduces the need for multiple credentials.
- *Managed key management systems* provide encryption and decryption services.
- *Compliance* to regulatory requirements such as Payment Card Industry (PCI), Health Insurance Portability and Accountability Act (HIPAA), Federal Risk and Authorization Management Program (FedRAMP), General Data Protection Action (GDPR), NIST-CSF, NIST 800-53, CSA-CCM, ISO27001, CCPA, and Clarifying Lawful Overseas Use of Data Act (CLOUD Act) are handled with cloud services or custom solutions.
- *Intrusion Detection System* and *Intrusion Prevention System*, if required, may be conducted in the cloud or on-premises.
- Data in transit is protected by using encryption at API end-points for the service APIs.

DevOps Hybrid Cloud

Figure 41—DevOps with Hybrid Cloud refers to DevOps-as-a-Service implemented over a hybrid cloud using a mix of on-premises, private cloud, and third-party public cloud computing services.

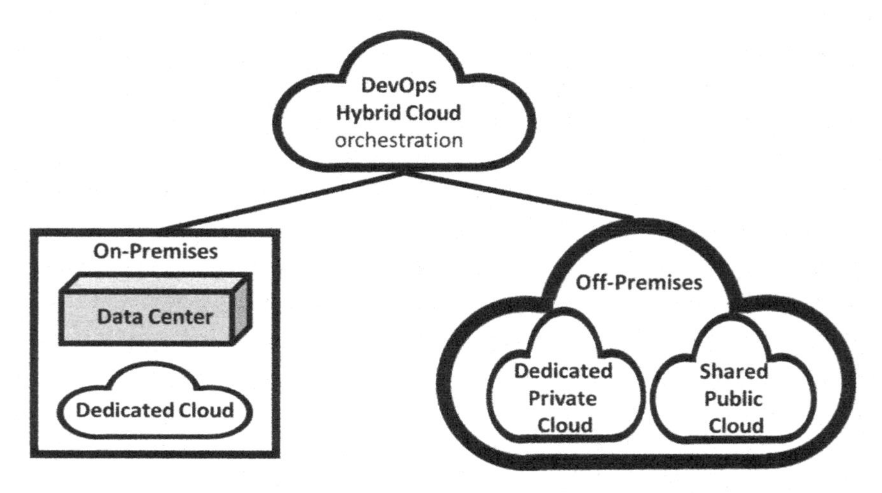

Figure 41—DevOps with Hybrid Cloud

DevOps *orchestration* tools operating together with hybrid cloud orchestration tools orchestrate dynamic workloads required for ephemeral and secure infrastructure needed for efficient and safe self-service DevOps applications.

A *DevOps Hybrid Cloud* enables converged self-service teams, tools, and processes with configurable and reusable pipelines that use resources spanning private and public cloud platforms.

Flexibility, performance, cost effectiveness, and scalability to support operational modes best suited for DevOps are features of well supported *Hybrid Clouds*. Public clouds offer scalability, while private clouds are better for handling sensitive data and offer higher I/O performance between tightly coupled services. Combined with *orchestration* and *automation*, *hybrid clouds* platforms are seamlessly integrated into a DevOps pipeline.

A well architected DevOps *hybrid cloud* orchestration stack, such as the one shown in **Figure 42—DevOps Hybrid Cloud Orchestration Stack**, unshackles IT from much of the everyday provisioning of traditional data center operations. In addition to eliminating manual processes, automation and orchestration also enhance security, reduce troubleshooting and improve an organization's compliance related to IT activities.

DevOps Hybrid Cloud Orchestration Tools

- Support plugins for DevOps pipeline orchestration tools.
- Support an abstraction layer such as Topology and Orchestration Specification for Cloud Applications (TOSCA).
- Enable abstract descriptions for resources (e.g., *VMs*, *containers*, *microservices*, and *databases*).
- Resolve most desirable deployment targets for resources based on factors such as cost, workload importance, performance and availability demand, and *security* requirements.
- Drive corresponding actions required to achieve the orchestrated outcome.
- *Hybrid cloud orchestration* platforms (e.g., Cloudify or Linux Foundation's OPEN-Orchestrator) are often based on TOSCA.

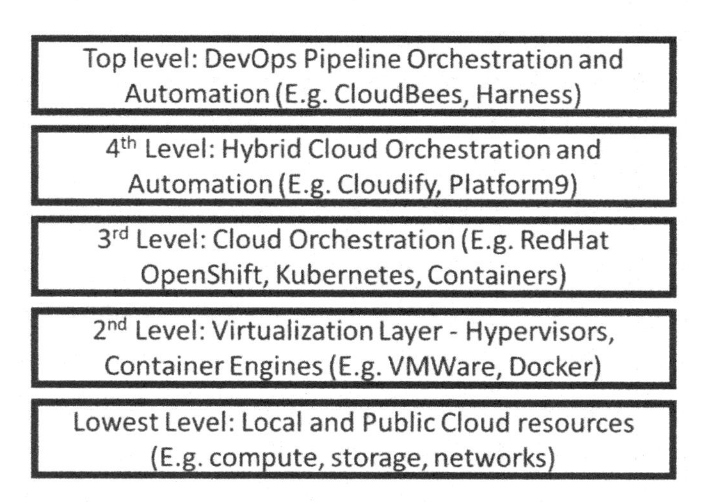

Figure 42—DevOps Hybrid Cloud Orchestration Stack

Examples of *DevOps Hybrid Cloud* recommended engineering practices are presented in the following list in two categories—engineering practices that apply to DevOps Hybrid Cloud and engineering practices that are peculiar to DevOps Hybrid Cloud:

Engineering practices that apply to DevOps Hybrid Cloud are as follows:

- *Hybrid Cloud* orchestration tools support plugins for DevOps pipeline orchestration tools.
- *Orchestration* addresses the provisioning of secure infrastructure services to support self-service DevOps applications, both ephemeral and persistent.
- *Hybrid Cloud* orchestration tools support an abstraction layer such as TOSCA.
- Orchestration can place loads based defined factors like cost, performance, availability, security, and other requirements.
- All communication of potentially sensitive data is encrypted.
- Orchestration instructions are created using API of the supported tools.
- Costing models are based on resource utilization to determine pricing of existing and proposed cloud environments.
- *Charge-back* and *show-back* visualization management tools are used to tell you which of your business units use the cloud infrastructure the most.
- Service dependencies are automatically discovered and move-group options are created as part of migration planning.
- Full-function testing is available for all service migrating between cloud to assure proper integration into supporting services before going live with a migration.
- Secrets management is provided to allow credential and *certificate management* across environments.
- *Governance* tool selection should ideally span environments and allow a common control plane as well as a home for deviation detection.
- *Governance* allows for business decision logic to be applied to workloads and cloud infrastructure.

- *Governance* follows a trusted framework, such as CIS/NIST as a baseline for operational controls.

Engineering practices that are peculiar to *DevOps Hybrid Cloud* are as follows:

- *DevOps-as-a-Service* is implemented over a hybrid cloud using a mix of on-premises, private cloud, and third-party public cloud computing services.
- **Hybrid Cloud orchestration** tools support abstract descriptions for resources (e.g., VMs, containers, microservices, and databases).
- **Hybrid cloud orchestration** can communicate with the required supplicant functions through the established API framework.
- DevOps *Hybrid Cloud Orchestration* stacks that are supported are defined.
- A dedicated reliable, high-speed/low-latency, redundant *WAN* connection to the public cloud is installed. Stakeholders have visibility into the deployment infrastructure in the desired private and public clouds.
- Tools help you transition workloads to the cloud by orchestrating virtual machine movement from a private environment to a public cloud.

DevOps Multi-Cloud

DevOps Multi-Cloud refers to *DevOps-as-a-Service* implemented over two or more cloud service providers, as shown in the **Figure 43—DevOps Multi-Cloud**. DevOps *orchestration* tools operating together with *multi-cloud orchestration* tools orchestrate dynamic workloads required for ephemeral and secure infrastructure needed for efficient and safe self-service DevOps applications.

A *DevOps Multi-Cloud* environment enables converged self-service teams, tools, and processes with configurable and reusable pipelines that use resources spanning multiple cloud platforms. Combined with orchestration and automation, multiple cloud platforms are seamlessly integrated into a DevOps pipeline.

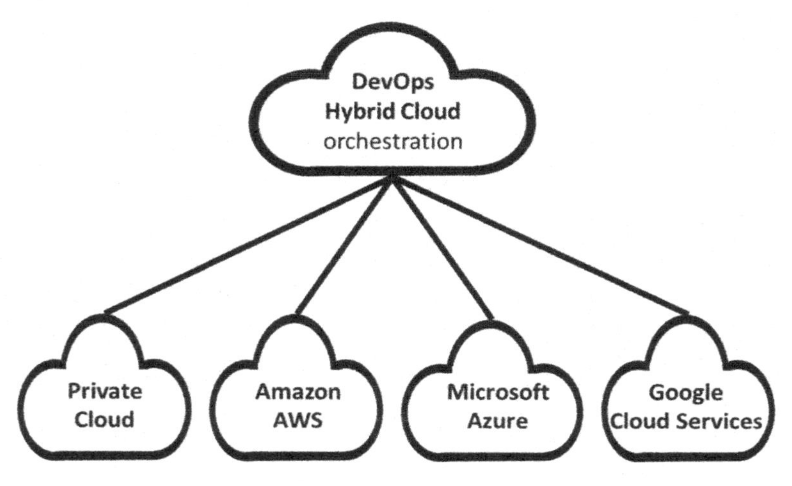

Figure 43—DevOps Multi-Cloud

The following capabilities are essential for *DevOps Multi-Cloud:*

- Ability to deploy to multiple targets and take advantage of the native capabilities (*Cloud-Native* applications) with DevOps processes and tools.
- DevOps tools adjust dynamically to cloud platform changes—in most cases, without human intervention.
- Logging, tagging, and other cloud *governance* and *security* concepts need to be considered in the tools, the applications, and the target cloud platforms to ensure consistency for stakeholders despite disparate capabilities of each cloud platform.
- *Multi-cloud* orchestration platforms (e.g., Cloudify or Linux Foundation's OPEN-Orchestrator) are often based on TOSCA,RS4 an open-source language used to describe the relationships and dependencies between services and applications that reside on a cloud-computing platform.

Applications being *Cloud-Native* is critical in a *Multi-Cloud* environment. Moving DevOps from a single or *Hybrid Cloud* to a Multi-Cloud world requires platforms and tools for capacity management and cost management. A globally aware traffic management strategy monitors infrastructure health across data centers and end-user experiences globally, while responding to control changes and system specifications at the speed of DevOps *Continuous Improvements.*

Examples of recommended engineering practices for *DevOps Multi-Cloud* are presented in the following list in two categories—engineering practices that apply to DevOps Multi-Cloud and engineering practices that are peculiar to DevOps Multi-Cloud.

Engineering practices that apply to DevOps Multi-Cloud are as follows:

- Stakeholders have visibility into what is deployed to each cloud.
- Stakeholders have visibility into utilization of workloads (CPU, memory, I/O) deployed for each *Cloud.*
- A *Software Version Management* tool (e.g., GitHub) is used to keep track of infrastructure code, including all the resources and variables needed to create the applications and pools of instances, virtual machines, or machines (each provider calls hosts by slightly different names).
- Operating system version packages are consistent across clouds.
- Each application deployed to the cloud is instrumented to support *Application Performance Monitoring (APM).*
- Applications behavior is verified to run consistently across virtualized instances when running the same operating system across the different clouds. Codify infrastructure is used by each application and then shared with all stakeholders in Dev, QA, and Ops to self-create and destroy them.
- Infrastructure used by each application is codified and then shared with all stakeholders in Dev, QA, Ops to scale infrastructure.
- IAC tools (e.g., Terraform) is used in layers to request resources from resource schedulers (e.g., Borg, Mesos, YARN, and Kubernetes) that schedule Docker containers, Hadoop, Spark, and many

other software tools sets up the physical infrastructure running resource schedulers as well as provisioning onto the scheduled grid.

- Visualize your IAC resources and gain insight into their dependencies.
- *Cloud-Native* advantages in each considered environment are known, understood, and utilized.
- Environments are constructed to follow well-architected frameworks. Bear in mind these change every year.
- Workload placement decisioning is re-evaluated regularly and in automated fashion.
- Traffic patterns adhere to data gravity and are egress sensitive (e.g., data may need to be *Multi-Cloud* hosted).

Engineering practices that are peculiar to DevOps Multi-Cloud are as follows:

- Documentation and training are provided to stakeholders on how to use and deploy workloads to *Multi-Cloud.*
- Infrastructure resources-as-code capability is used to make the application environment as similar as possible across all *Clouds* (e.g., Terraform).
- To support resiliency, applications are designed to run on any of the cloud providers, and the application can use any of the available providers to satisfy a given request. If one provider is unavailable, the other provider can take over to process requests for the application.
- Applications code that is required to run and perform equally on any *Cloud* is limited to using capabilities common to all the *Cloud* providers.
- Multi-cloud quality assurance practices are employed to ensure applications, which tend to be more complex when they are designed to work across multiple *Cloud* providers, do not lead to an increase in errors and other problems.

- *Cloud-agnostic* coding tools are used to codify the configuration for software defined networks to automatically setup and modify settings by interfacing with the control layer (e.g., HashiCorp HCL).
- *Cloud-agnostic* capabilities are used to manage multiple providers and handle cross-cloud dependencies, simplifying *Multi-Cloud* management and orchestration and helping operators build large-scale *Multi-Cloud* infrastructures.
- A *Registry* is used by stakeholders to publish, version, and share common provisioning modules with the *Multi-Cloud* community (e.g., Terraform Registry).
- There is cross-cloud secrets management.
- There is cross-cloud health monitoring at the service and application levels.
- There is cross-cloud governance and security decisioning made at a level of abstraction above the tactical layer.

DevOps Multi-Cloud Services

As indicated in **Figure 44—Multi-Cloud Services Forecast**, the future of DevOps is mastery of Multi-Cloud environments and services. *Multi-Cloud* provides the ultimate DevOps infrastructure that supports the widest, most flexible range of application deployments requirements. Enterprises that assemble harmonized *Multi-Cloud* platforms are positioned for competitive advantage and lower costs.

As indicated by **Figure 45—Cloud Service Providers Comparison**, managing *Multi-Cloud* environments can be more complicated than managing a single *Cloud* environment.[RW65, RW66]

- It is not a "write once, run everywhere" situation.
- CSPs have their own feature sets, *Graphical User Interfaces (GUIs)*, and access protocols.
- A *Multi-Cloud* management system must do the work to provide deep integration with each CSP.

- A success strategy considers challenges, benefits and a clear path to successful Multi-Cloud management.

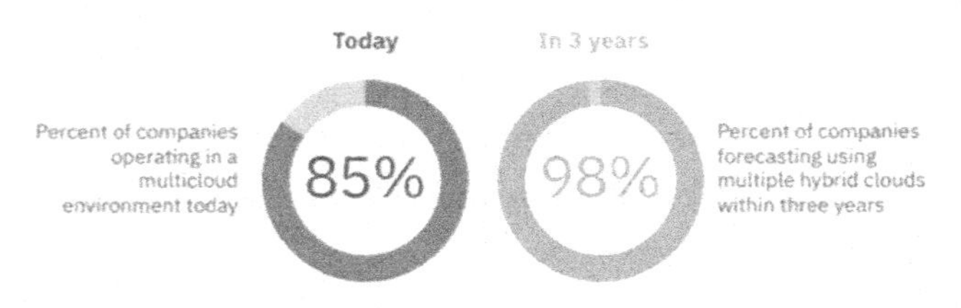

Figure 44—Multi-Cloud Services Forecast

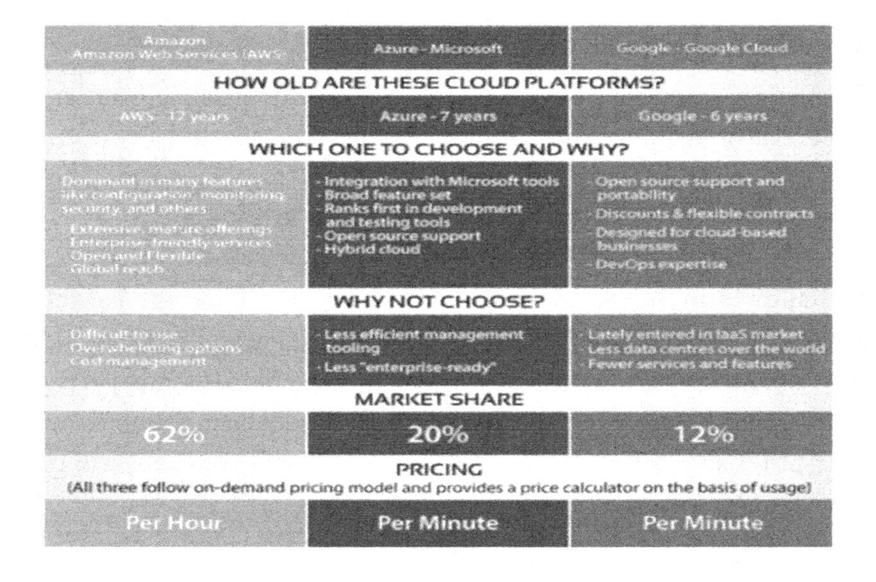

Figure 45—Cloud Service Providers Comparison

As indicated by the **Figure 46—Cloud Services Comparison**, *Multi-Cloud* is essential to enterprise DevOps strategy because of wide-ranging advantages, including risk mitigation; avoiding vendor lock-in; optimization; and unique capabilities for AI, IOT and ML. DevOps is mandatory for *Multi-Cloud* deployments.

PRODUCT	aws	Microsoft Azure	Google Cloud Platform
Virtual Servers	Instances	VMs	VM Instances
Platform-as-a-Service	Elastic Beanstalk	Cloud Services	App Engine
Serverless Computing	Lambda	Azure Functions	Cloud Functions
Docker Management	ECS	Container Service	Container Engine
Kubernetes Management	EKS	Kubernetes Service	Kubernetes Engine
Object Storage	S3	Block Blob	Cloud Storage
Archive Storage	Glacier	Archive Storage	Coldline
File Storage	EFS	Azure Files	ZFS / Avere
Global Content Delivery	CloudFront	Delivery Network	Cloud CDN
Managed Data Warehouse	Redshift	SQL Warehouse	Big Query

Figure 46—Cloud Services Comparison

As indicated by **Figure 47—Multi-Cloud Architecture,**[RW67] DevOps systems must evolve towards mastery of *Multi-Cloud* ecosystems, new platforms, and tools. DevOps must deploy, connect, maintain, and scale *containerized microservices* and cloud-native applications so they're portable and easily managed and orchestrated. As illustrated by **Figure 48—Traditional Cloud-Native Apps,**[RW67] *Cloud-Native* apps are modular and service-oriented; comprised of collections of containers and *microservices*; and based on a scaled-out architecture that is easier to automate, move, and scale. A common *container* environment helps mitigate CSP differences by isolating engineers from the nuances of *VM* management; however, issues of network, infrastructure, cost optimization, security, and availability are *CSP*-specific. A globally aware traffic management strategy monitors infrastructure health across *data centers* and end-user experiences globally, while responding to control changes and system specifications at the speed of DevOps.

The following are recommended engineering practices for DevOps *Multi-Cloud* Services presented in two categories—engineering practices that apply to DevOps Multi-Cloud Services and engineering practices that are peculiar to DevOps Multi-Cloud Services.

Engineering practices that apply to DevOps Multi-Cloud Services are as follows:

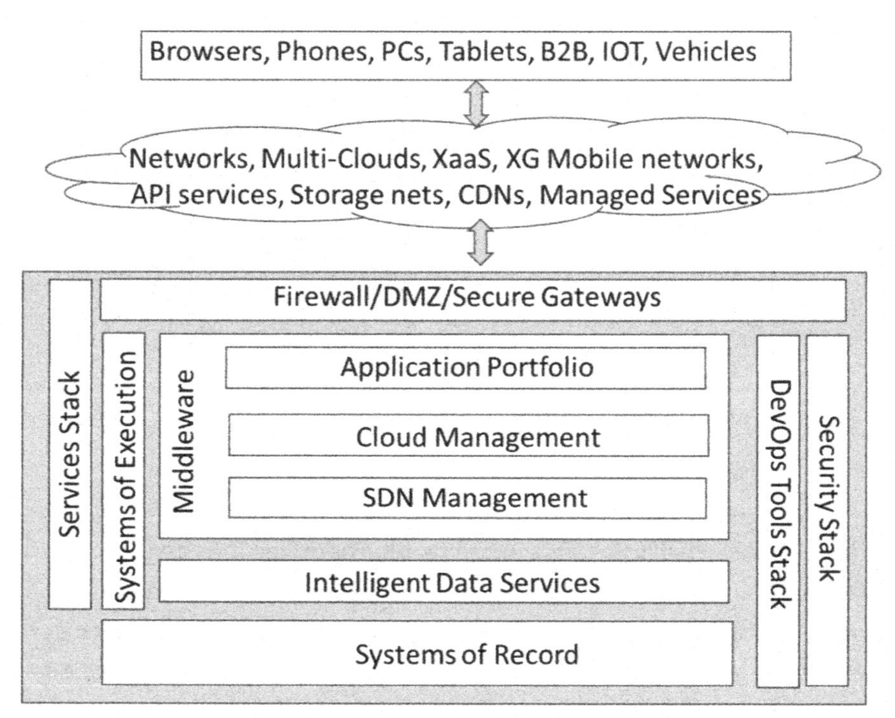

Figure 47—Multi-Cloud Architecture

- DevOps tools reliably deploy code to public, private, and *Hybrid Clouds.*
- Automated platform tests executed prior to deployment of new software or system changes verify that the same code coupled with data performs as expected on each target platform.
- *Containers* are integrated into your DevOps toolchain to support application portability, *orchestration*, and management.
- Applications are designed using *Cloud-Native* and *twelve-factor app* design concepts to ensure they can be deployed across the *multi-cloud.*

- A common *container* environment helps to reduce *Multi-cloud* application deployment issues by isolating engineers from the nuances of *virtual machine* management.
- The same *container* is used for all environments. For example, config files are mapped via *Docker* volumes, and environment variables configure applications dynamically to suit the deployment environment.
- *Container Orchestration* tools are used to manage clusters with the following capabilities: provisioning, monitoring, rolling upgrades and roll-back, configuration-as-text, policies for placement, scalability, and administration.
- Policies ensure that applications are matched to the *Cloud* services that best match their strengths.
- Business policies considering security, compliance, insurance, engineering costs, maintenance and ROI factors determine when to reengineer an application into cloud-native and migrate it to a *Cloud* service.
- A globally aware traffic management strategy monitors infrastructure health across data centers and end-user experiences globally to facilitate responding to control changes at the speed of DevOps.
- User activity is monitored and analyzed to understand how applications and systems are being used and which processes are the biggest productivity boosters.

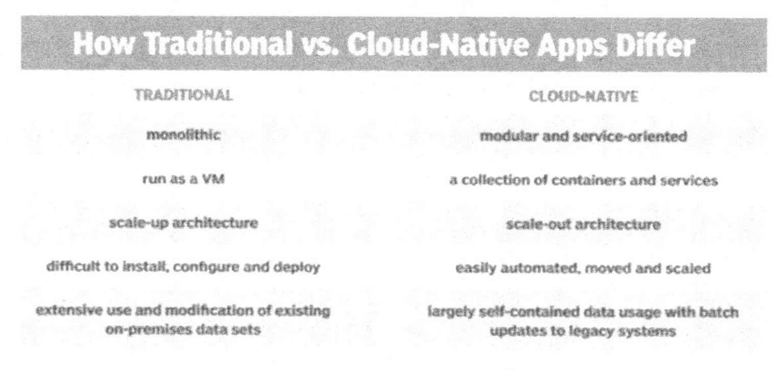

How Traditional vs. Cloud-Native Apps Differ

TRADITIONAL	CLOUD-NATIVE
monolithic	modular and service-oriented
run as a VM	a collection of containers and services
scale-up architecture	scale-out architecture
difficult to install, configure and deploy	easily automated, moved and scaled
extensive use and modification of existing on-premises data sets	largely self-contained data usage with batch updates to legacy systems

Figure 48—Traditional Cloud-Native Apps [RW67]

Engineering practices that are peculiar to DevOps Multi-Cloud Services are as follows:

- A *Multi-Cloud* architecture team with key stakeholders from Dev, Infra, Ops, Security, and users determine *Multi-Cloud* goals, architecture, services, and policies.
- Realistic *Multi-Cloud* goals are set including rules for choosing which platform works best for each application. *Multi-Cloud* goals are reviewed periodically to account for the evolving cloud landscape.
- *Multi-Cloud* networks security strategy includes securing the inside of the network, not just the perimeter, because you no longer control the network perimeter.
- A *Multi-Cloud* management platform accesses and consolidates data from all sources to provide a *Multi-Cloud* perspective.
- A common set of metrics provides visibility of your *Multi-Cloud* environments.
- Policy-driven automation resolves concerns about employees failing to comply with *Multi-Cloud* security policies.
- DevOps processes are designed and, when necessary, re-engineered for *Multi-Cloud* environments.
- Automated tests verify APIs and user interfaces for each *Cloud* service perform as expected each time a *Cloud* service is changed.
- A single standard multi-cloud deployment policy is applied automatically to each environment, (covering such areas as virtual servers, workloads, data storage, traffic flows, *compliance* and *regulation*, and reporting and *security*) to ensure updates and changes propagate seamlessly from environment to environment.
- A custom combination of *Multi-Cloud* services, with shared policies and configurations (e.g., comparable read/write loads, data transfer needs, and network latency), are employed to avoid vendor lock-in. All exceptions require justification and exception approval.
- Costs are proactively monitored, and cost controls are adjusted across multiple *CSPs* to ensure cost-effective alternatives are being used.

- User and admin training is offered for both DevOps and *Cloud* computing.

Five Levels of Infrastructure Maturity

As indicated in prior sections of this chapter, DevOps needs elastic infrastructure to work best. The more elastic is, the better or more mature it is from the point of view of DevOps capabilities. DevOps elastic infrastructures maturity levels are shown in **Figure 49—Elastic Infrastructure Maturity Levels**.

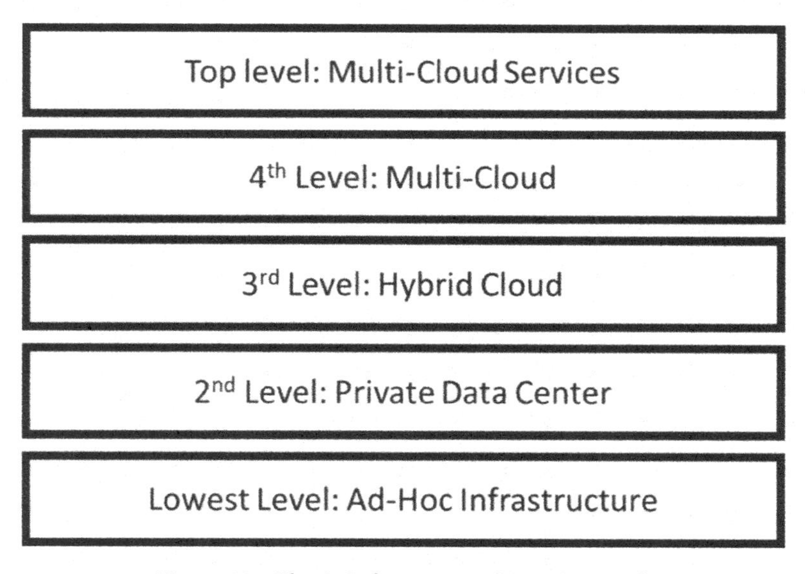

Figure 49—Elastic Infrastructure Maturity Levels

In this chapter, the various infrastructures and levels of maturity of infrastructures that are most effective for engineering DevOps were discussed. This chapter, combined with the prior two chapters covering applications and pipelines, complete three axes of DevOps engineering that can mature independently but are most effective when they mature together. The next chapter explains how applications, pipelines, and infrastructures are used together for testing application changes over infrastructures as the changes traverse the pipeline during an application change release process.

16

Continuous Test Engineering

In my DevOps consulting experience, in both enterprise and manufacturing environments, I see many value streams in which test activities such as finding and setting up test environments, test configurations, and test results interpretation are a primary source of bottlenecks. Success with DevOps requires that these bottlenecks are reduced. A thorough understanding and disciplined implementation of *Continuous Testing* and related recommended engineering practices is required.[RW58]

!! Key Concept !! Continuous Testing Is Essential for DevOps

Continuous Test Engineering is a quality assessment strategy in which most tests are automated and integrated as a core and essential part of DevOps. Continuous testing is much more than simply "automating tests."

Why Is Continuous Test Engineering Important to DevOps?

A comprehensive set of tests are, in aggregate, a prescription of expected behaviors for applications within the production environments. The expected behaviors are more than application features and functions. Non-functional requirements such as performance, maintainability, and reliability are essential. *Continuous tests* are designed to cover both functional and *Non-Functional Requirements (NFRs)*. Capabilities needed to operate an application in each of its deployment environments are part of *continuous testing*.

How Is Continuous Testing Engineered for DevOps?

Figure 50—Continuous Test Engineering Blueprint shows that with *continuous testing*, as many relevant tests as possible are executed as early as possible in the pipeline.[RW60] Assessments are performed "continuously" on incremental product changes using a production equivalent environment (Physical and/or Virtual), which are orchestrated for each of the end-to-end pipeline stages from Dev through Integration, Pre-Prod and Production stages. Tests are selected automatically at each stage and triggered to run automatically by the toolchain prior to exiting each stage of the pipeline. Results of tests are a critical factor for deciding whether an application change is ready for promotion from one stage to the next stage in the pipeline or rejected and scheduled for remediation. The aggregate set of test results accumulated over the pipeline provide important data to decide whether a change is acceptable for release to production.

The following are example assessments for each stage of the pipeline:

Plan Stage Continuous Test Assessment Examples

- Test plans tasks are in backlog for each user story
- Test resources are identified for each test task (physical and virtual)

Dev Stage Continuous Test Assessment Examples

- Unit and static analysis test verdicts for each coding task
- Functional, not functional, and regression test verdicts for each user story

Integration Stage Continuous Test Assessment Examples

- Dev test results are verified for each commit
- Integration test verdicts

Pre-Prod Stage Continuous Test Assessment Examples

- System and User Acceptance test verdicts
- Verdicts of deployment tests

Post-Prod Stage Continuous Testing Assessments Examples

- A/B application comparison test statistics
- Test escapes—production failures for which there is no defined test case

Continuous testing engineering follows five tenets: shift left, fail early, fail often, test fast, and be relevant.

!! Key Concept !! Continuous Test Engineering Tenets

Shift Left: Conduct each test as early in the pipeline as possible.

Fail Early: Arrange the tests so that the most likely problems are found in the earliest possible stage in the DevOps pipeline.

Fail Often: Run tests frequently and with many different conditions.

Test Fast: Arrange tests to run in quick cycles.

Be Relevant: Focus on the most important tests and results.

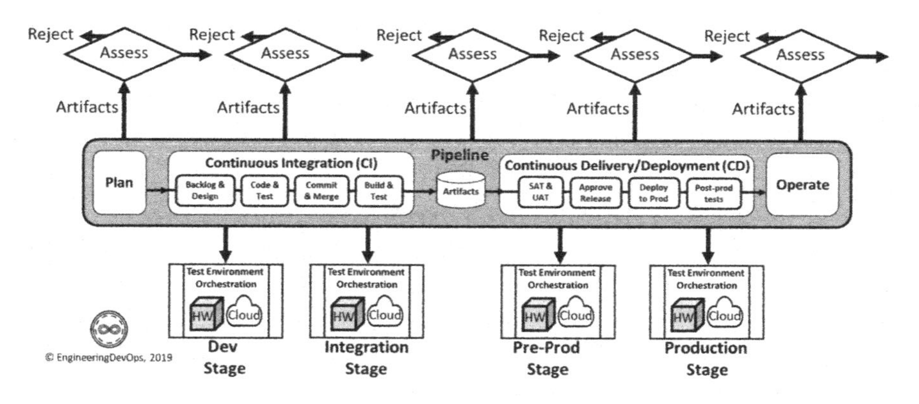

Figure 50—Continuous Test Engineering Blueprint

These tenets and the following recommended engineering practices for *continuous* test engineering are essential to success with DevOps:

- Test plans are reviewed by product developers, product architects, test engineers, operations staff, and tools and infrastructure staff to ensure the plan has sufficient test coverage to meet customer, product, and operation requirements and to ensure tools and infrastructure requirements are planned.

- *Release* performance tests and other non-functional tests are automated to verify that no unacceptable degradations are released.
- At least 85% of release regression tests are automated.
- Product-level test plans provide overall guidance for product test strategies and policies and indicate the strategy/contract for measuring test coverage.
- New unit and functional regression tests that are necessary to test a software change are created together with the code and integrated into the trunk branch at the same time the code is. The new tests are then used to test the code after integration.
- A test script standard is used to guide test script creation to ensure the scripts are performing the intended test purpose and are maintainable.
- Development changes are *"Pre-Flight"* tested in a clone of the production environment prior to being integrated to the trunk branch. (Note: "production environment" means "variations of customer configurations of a product.")
- Tests are triggered automatically during each stage of the *Continuous Delivery* pipeline.
- Test reports include verdicts of each test (pass/fail/inconclusive) and any logs from the system under test or test tools, time-correlated to test steps.
- Test tools, test plans, and tests are version managed.
- The version of test tools, test plans, and tests can be reinstated when there is a need to revert to a prior version of an application.
- Resources required for testing are provisioned and orchestrated automatically when a test is triggered. These can be physical resources (e.g., servers, devices, etc.) when testing embedded systems, IOT or network devices, or virtual resources (e.g., images in containers, VMs) when testing enterprise software, and topologies consisting of a mixture of physical and virtual resources (e.g., network testing).

Advanced Continuous Test Engineering

As indicated in **Figure 51—Advanced Continuous Test Squeeze**, more mature DevOps implementations accelerate the speed and frequency of continuous delivery pipelines and releases.[RW59] The volume of tests to be run increases to the point that basic *Continuous Flow* test practices become a bottleneck.RW69 More advanced continuous test engineering practices become essential. Consequences of not using advanced *continuous test* engineering methods include the following:

- Quality problems propagate through the pipeline to production.
- Customer-reported failures increase.
- Non-value-added work to fix problems increases.
- Internal costs increase.

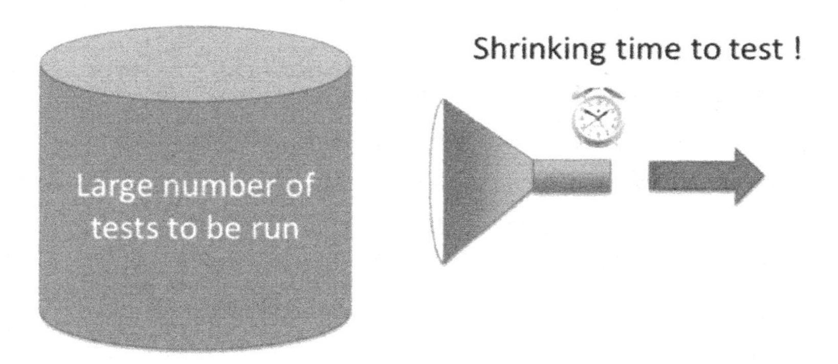

Figure 51—Advanced Continuous Test Squeeze

Intelligent test *tool frameworks* and test tools that are well integrated into the DevOps *toolchain* and workflows are essential to accelerate testing at scale.[RW72] **Figure 52—Advanced Continuous Test Engineering Blueprint** illustrates an advanced continuous test engineering blueprint that is designed to accelerate test creation, test management, test orchestration, test execution, and test results analysis and reporting and in-production testing.

The following steps are recommended practices for implementing advanced continuous test engineering:

- Do a value stream analysis from the *point-of-view* of testing to identify bottlenecks for test creation, test orchestration, test execution, and reporting.
- Select a test automation framework that can orchestrate and monitor test results for the end-to-end pipeline, from test creation through to operations.
- Determine and implement metrics for each pipeline stage. Select tools to address bottlenecks.
- Integrate the test tools into the test tool framework.
- Implement *Blue/Green* deployments to verify the solution safely.
- Implement *A/B* testing with the applications and test frameworks.
- Scale up the tests using advanced test creation methods.
- Speed up the tests using advanced test selection and analysis methods.

In this chapter, it was explained that Continuous Testing engineering requires a strategic approach. It is much more than simply the automation of tests. Continuous Test engineering requires a combination of orchestration of elastic infrastructures, combined with tests that are targeted to application modules and the integration of test tools with tests across the pipeline. These engineering practices are critical for even *The First Way* of DevOps because testing provides the assessment data required to judge whether an application change is ready to be promoted through each pipeline stage to release. The next chapter will explain that Continuous Monitoring, when engineered according to a solid engineering blueprint following engineering practices, works together with Continuous Testing, application, infrastructures, and pipelines to provide visibility needed for maturing DevOps towards *The Second Way of DevOps—Continuous Feedback.*

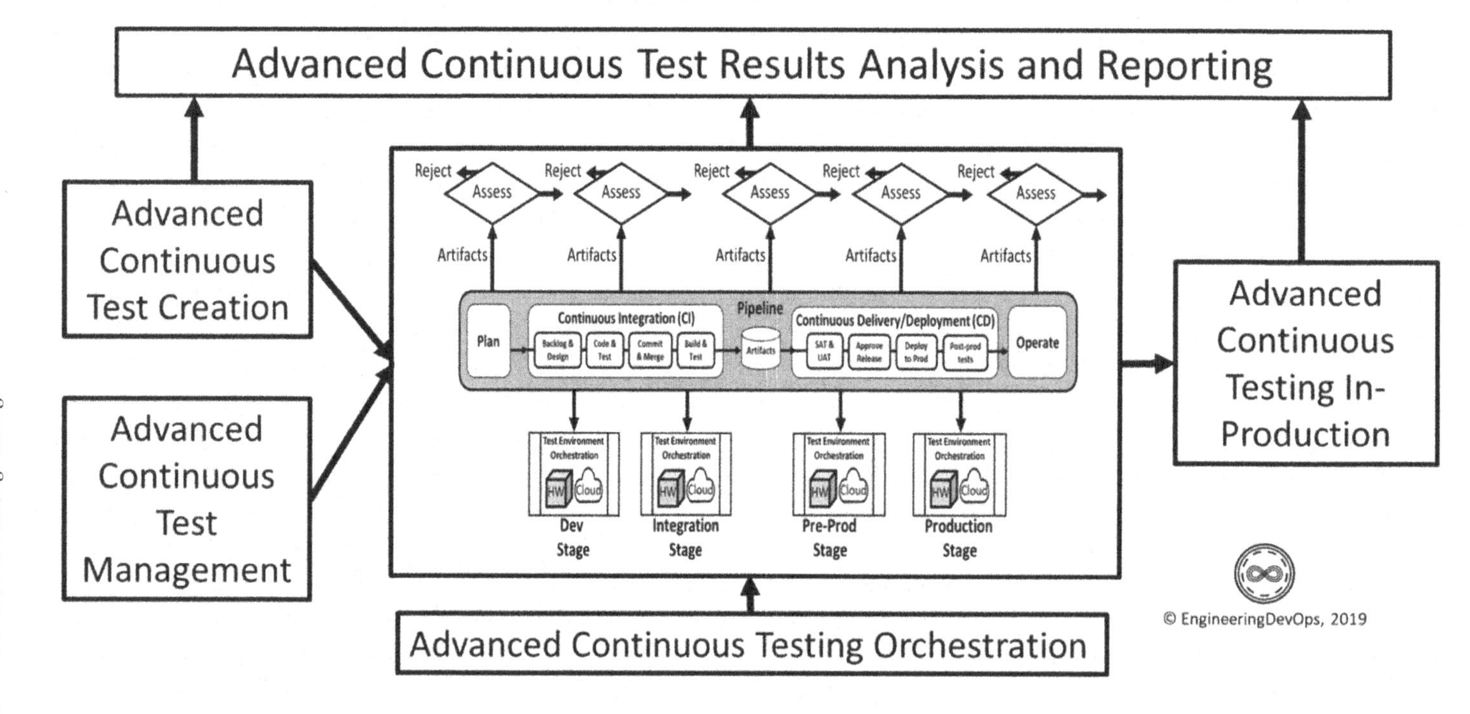

Figure 52—Advanced Continuous Test Engineering Blueprint

17

Continuous Monitoring Engineering

Continuous Monitoring provides visibility of the health of all things that are important to the operation and performance of the application, the *Continuous Delivery* pipeline, and the infrastructure that they depend on.

The health of all Nine Pillars of DevOps affects the performance of *Continuous Delivery*, including leadership and culture. *Continuous Monitoring* is an essential tool that enables collaboration across IT specialties, reducing or eliminating the finger-pointing and other unproductive behaviors that show up all too often in organizations with operational silos. *Continuous Monitoring* is different than conventional monitoring. It requires taking a holistic view of the complex application environment. *Continuous Monitoring* requires organizational alignment. Development, QA/Test, and Operations must agree upon an implementation strategy that allows *Continuous Monitoring* to become a source of truth for the entire organization.

Why Is Continuous Monitoring Engineering Important to DevOps?

With DevOps, the rate of request for change increases, and with that comes an increase of risk. Continuous Monitoring provides a means to control the risk and provide a feedback mechanism for the health of the application release, the pipeline, and the infrastructure. The following are example benefits of well-engineered *Continuous Monitoring*:

- End-to-end visibility into your applications and database performance
- Continuous feedback loops to support continuous improvements
- *Continuous Monitoring* of the *End-User-Experience (EUE)* providing customer feedback essential to identifying product preferences and improvements
- Early detection of performance bottlenecks across the application, pipeline, and infrastructure
- Higher confidence in achieving *SLAs* in production
- Better prediction of code behavior
- Risk management
- Cost management
- Performance management
- Application complexity management
- Better management of the Digital Experience

How Is Continuous Monitoring Engineered for DevOps?

As illustrated in **Figure 53—Continuous Monitoring Engineering Blueprint** a key challenge for *Continuous Monitoring* Engineering is the selection of data to be monitored, how to aggregate it, and how to implement decisions based on the aggregated data.

The *Continuous Monitoring* Data Layer includes, for example, the following:

- Log event streams for asynchronous and synchronous metrics needed to assess operational performance deviations from applications, pipelines, and infrastructure components (Notifications intended to be read by a human are pushed to a ticket queue, email, or pager. Alert systems may offer users subscription services to tailor their alerts to their specific roles. Examples of alert information include reactive alerts—"Something is broken!"; predictive alerts—"Something is going to break!"; and informative alerts—"What was happening at the time of an alert?")

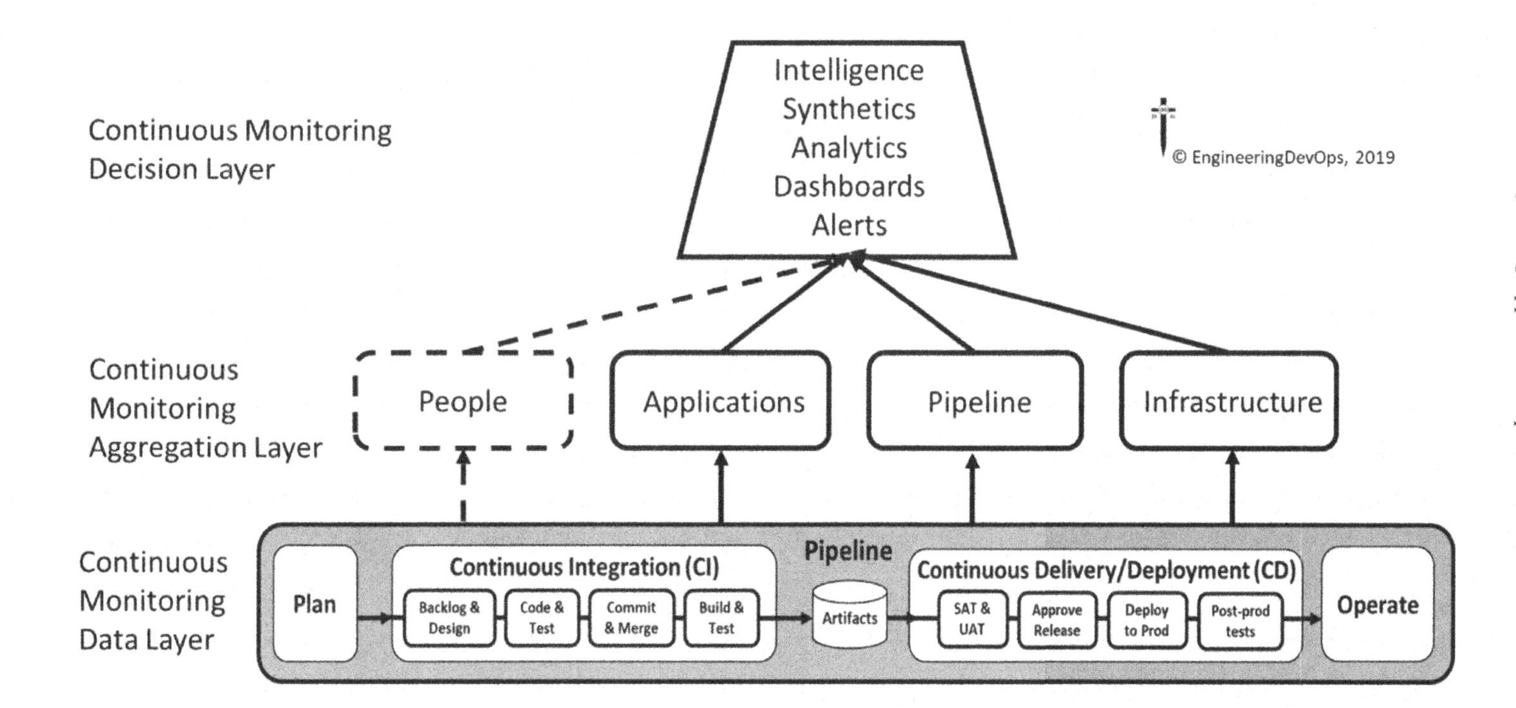

Figure 53—Continuous Monitoring Engineering Blueprint

- Health checks for individual components
- *Continuous Monitoring* agents positioned with the deployed code act as event collectors to send to an event aggregation server

The *Continuous Monitoring* Aggregation Layer includes, for example, the following:

- Aggregation, deduplication, and prioritization of logged event streams
- Ability to span multiple environments (i.e., Cloud, Hybrid-Cloud, Multi-Cloud, On-Premise)

The *Continuous Monitoring* Decision Layer includes, for example, the following:

- Alerts that are derived from aggregated log event streams
- Displays and dashboards, usually web-based, providing a visual summary built to expose metrics that are most important to users (*Dashboards* may include selectors, filters, and telescoping features that allow users to refine the view. Examples of items to display include current values, trend graphs, queue lengths, priority, and owners.)
- *Release* comparison giving clear insight into release performance
- Synthetic transactions, which can be early detectors of performance problems

Processes, procedures, and policies that govern "how" you will operate and "who" will operate critical functions must be defined. This requires aligning your business goals across your entire organization to ensure that Continuous Integration and delivery are achieved.

DevOps requires collaborative teams. It is perhaps the most critical success factor for high-performance DevOps. Yet collaboration is often overlooked when engineering monitoring. Monitoring people in a constructive positive way is critical. Metrics that emphasize reinforcement of positive behaviors instead of punitive metrics are most recommended. Some of the things that can be monitored are listed as follows.

1. Work backlog trends
2. Percent of commits successful
3. Percent of work that is creative versus corrective
4. Percent of *SRE* time in development
5. Job satisfaction ratings

Many application performance bottlenecks originate in the *database*, but too often stakeholders have little or no visibility into database performance. Here are some items that should be monitored for databases:

- Database size
- Alert when database thresholds are violated
- Trend resource consumption, database objects, schema statistics
- Database growth
- Active user counts
- Response time
- Calls and errors

The following are examples of popular tools used for *Continuous Monitoring* solutions:

- Intelligent analytics: Tableau
- Value stream analytics: Plutora, CollabNet, VersionOne, Tasktop
- Application release metrics and analytics: CloudBees, ElectricFlow, XebiaLabs, CA Technologies, Azure DevOps
- Dashboard tools: Prometheus, Hygieia (Capital One), DataDog, Service Now
- Application performance monitoring tools: Dynatrace, App Dynamics, New Relic
- Infrastructure monitoring, logs, and alert tools: Nagios, Splunk, Riverbed

The following are examples of recommended engineering practices for engineering *Continuous Monitoring* presented in three categories: Engineering Continuous Monitoring for Applications, Engineering Continuous Monitoring for Pipelines, and Engineering Continuous Monitoring for Infrastructures.

Engineering Continuous Monitoring for Applications

- An effective continuous monitoring strategy keeps track of the health of all things that are most important to the operation and performance of the continuous delivery pipeline, including team performance, application and database performance, integration, testing, delivery, deployment, and infrastructures.
- A well-engineered Continuous Monitoring solution requires the following:
    - Balance device- and traffic-oriented metrics with more qualitative *End-User-Experience (EUE)*, *(MOS/R-Factor)* metrics.
    - Focus action-oriented performance metrics that can anticipate demand.
    - Measure end-user experience for complex apps, web and mobile services, SaaS, outsourced and hosted services, homegrown apps, enterprise apps, and legacy and back-office systems.
- Goals, metrics, and thresholds are set for application, database, and DevOps infrastructure performance.
- Logging and proactive alert systems include application performance data to make it easy to detect and correct application performance failures or degradations.
- Monitor outliers, not just averages.
- As with any distributed system, monitoring microservices runtime stats are required to know the application performance and automate corrective actions. Measure requests per second, avail-

able memory, number of threads, number of connections, failed authentications, and expired tokens. Use reactive microservices monitoring to detect problems, trigger circuit breakers, detect heavy loads, and spin up new instances with the cloud orchestration platform.

Engineering Continuous Monitoring for Pipelines

- Deployment and release decisions for application changes include thresholds that include performance metrics.
- Deployment and release decisions for database changes include thresholds that include performance metrics. It is not a test of the database itself but rather the performance of the database contribution to app performance.
- Logging and proactive alert systems include database performance data to make it easy to detect and correct database performance failures or degradations.
- Logs and proactive system alerts for application and database performance exceptions are organized in a manner to quickly identify the highest-priority problems.
- Snapshot and trend results of application and database performance metrics are collected for each DevOps pipeline stage (e.g., Dev, CI, Stage, etc.) and made visible to everyone in the Development, QA, and Ops Teams.
- Every integration failure indicates a problem has been found that may otherwise have made it into production. Common integration metrics include: build success rate, unit test pass rate, static analysis exceptions, integration test coverage, adherence to coding standards, cyclomatic complexity, and smoke test results.

Engineering Continuous Monitoring for Infrastructures

- Logging and proactive alert systems include infrastructure performance data to make it easy to detect and correct application performance failures or degradations.
- Application and database resource consumption (e.g., memory, response time, process time, etc.) is reported so that users can see what application components are consuming and would be beneficial to optimize.
- Key Performance Indicators (KPIs) for the DevOps infrastructure components include performance metrics that are automatically gathered, calculated, and made visible to anyone on the team that subscribes to them.

Continuous Monitoring systems can be classified into four levels of maturity listed as follows, with each level building on the lower levels:

1. Optimizing—self-correcting
2. Proactive—integration of *Continuous Monitoring* sources
3. Reactive—separate *Continuous Monitoring* sources
4. Chaos—missing *Continuous Monitoring* sources

In this chapter, the Engineering Continuous Monitoring blueprint and engineering practices that are needed to provide visibility into DevOps Applications, Pipelines, and Infrastructures were explained. Without visibility, it is not possible to mature *The Second Way* of DevOps. It is also not recommended to automate continuous deployment until well-engineered monitoring and feedback tools and practices are in place.

In the next chapter, Continuous Delivery and Deployment engineering blueprints and associated recommended engineering practices are explained.

18

Continuous Delivery and Deployment Engineering

When explaining **CD** for DevOps I am reminded of the classic Abbot and Costello comedy routine, "Who's on First?" Yes. Who? Who is on first? I am asking you! That's why I told you. Who? Yes!

The abbreviation "**CD**" is ambiguous because it is often used to refer to the "Continuous Delivery pipeline," "Continuous Delivery," and "Continuous Deployment," which are related but three separate things.

According to Jez Humble, ***Continuous Delivery*** is the ***ability*** to get changes into production or into the hands of users safely and quickly in a sustainable way.[RB6] A key distinction is that *Continuous Delivery* is about the ***ability*** to get changes into production but does not necessarily include deployment to production. There are a lot of cases, such as new platform products, in which it is not desirable to deploy all software releases to production, even though, from an internal process point of view, there are many reasons to get the software release ready to deploy to production.

Continuous Deployment is a set of practices that enable every change that passes automated tests to be ***automatically deployed to production***. Continuous Deployment takes continuous delivery to a higher level of automation. Continuous Deployment depends on Continuous Delivery but takes the changes all the way to production.

Continuous Delivery and *Deployment* both depend on ***Continuous Delivery pipelines*** that continuously integrate software developed by the development team, build release candidate artifacts, and run automated tests on those artifacts to detect problems. The best practice for CD pipelines requires pushing artifacts into increasingly production-like environments to minimize the chance that software may fail when deployed to the production environment.

You have engineered *Continuous Delivery* properly when the following occurs:

- Your software is deployable throughout its life cycle.
- Your team prioritizes keeping the software deployable over working on new features.
- Anybody can get fast, automated feedback on the production readiness of their systems any time somebody makes a change to them.
- You can perform push-button deployments of any version of the software to any environment on-demand.

You have engineered *Continuous Deployment* properly when the following occurs:

- Your software is automatically deployed after continuous delivery.

Hey, don't blame me for the ambiguity around the different uses of the word deployment when used with DevOps! Oh—I should mention one additional common use of the word "deployment" with DevOps, because it can be really confusing. The *orchestration*, instantiation, and release of applications over infrastructure for Dev, Integration and Pre-Production stages, for testing purposes prior to deployment to production, is also called deployment. So be careful to understand and clarify the context of the word "deployment" when used with DevOps!

Why Is Continuous Delivery and Deployment Important to Engineering DevOps?

Benefits of well-engineered *Continuous Delivery* and *Deployment* include the following:

- A well-engineered *Continuous Delivery* pipeline provides visibility for software as it propagates through each stage.
- Quick lead times and more frequent releases provide quicker access to user feedback.
- Smaller, incremental, controlled *releases* are less painful, and failure events are lower risk.
- Reduced lead improves time-to-market for innovative new features.
- Software quality and stability are improved when each change is following a disciplined, well-engineered *Continuous Delivery* pipeline.
- Cost of software change is reduced when lean engineering practices are applied to the *Continuous Delivery* pipeline.
- Customer and employee satisfaction improves when they see positive results of the *Continuous Delivery* practices.

Well-engineered *Continuous Delivery* and *Deployment* avoids the following types of problems:

- Inefficient Change Review Board meetings required to approve releases
- Voluminous release documentation instead of automated deployments
- Reliance on manual error prone testing
- Frequent corrections to the manual release process
- Manual configuration errors
- Lengthy manual deployments
- Frequent release roll-backs caused by manual errors
- Unexpected interruptions during a long release
- Sitting bleary-eyed in front of a monitor during release hell nights and weekends

How Is Continuous Delivery and Deployment Engineered?

Figure 54—Continuous Delivery and Deployment Engineering Blueprint illustrates a CD pipeline that is typical of well-engineered mature DevOps implementations. The CD pipeline starts with *release* candidate *artifacts* delivered from the CI pipeline into an *artifacts repository*. The CD pipeline is automated, and artifacts are deployed to a pre-production staging environment for system and user acceptance testing. These tests are automated as much as possible to minimize processing delays and bottlenecks. Once the tests pass, the artifacts are assessed with release polices to win approval for deployment. The release acceptance criteria determined by a *Change Approval Board (CAB)* are implemented as codified policies as much as possible. Once approved, the artifacts are deployed to production according to a deployment schedule. To minimize risk, the deployments are conducted using the safe *Green/Blue* and *Dark Launch* methods that are explained in this section. Once deployed, there may be additional post-deployment tests to evaluate software options using *A/B testing* and *Canary* testing methods.

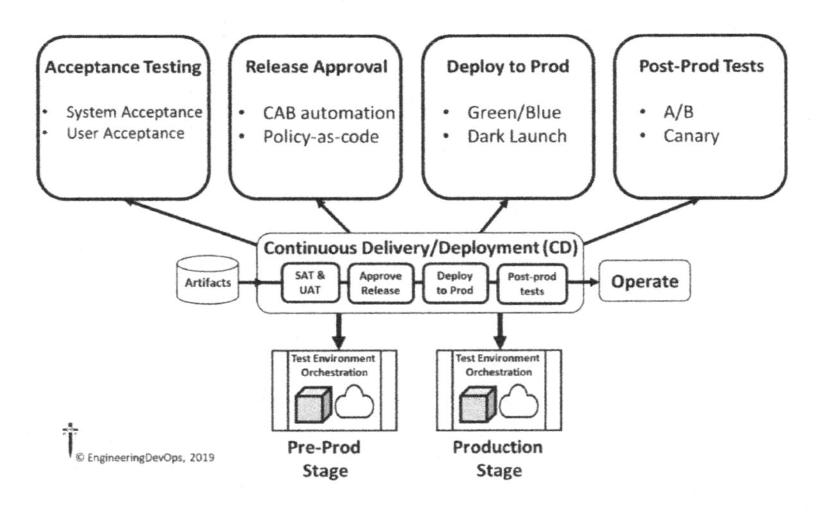

Figure 54—Continuous Delivery and Deployment Engineering Blueprint

Continuous Delivery uses the following tenets:

- **Automate the build, deploy, test, and release process**. This ensures consistent configuration of the system, environments, and the release process. This enables your processes to be less error-prone, more traceable, and easier to control, and it will help you remediate problems. Overall, you get a better-quality result.
- **Frequent small releases are safer**. Smaller, frequent releases reduce the risk associated with bigger, less frequent releases. It's much easier to roll back the application and its associated configuration (including its environment, deployment process, and data).
- **Fast feedback is essential**. Any change, of whatever kind, needs to trigger the feedback process. The feedback must be delivered as soon as possible. The delivery team must receive feedback and act.

For Continuous DevOps SREs, Dev and business stakeholders need to agree to a release strategy that defines the following components:

- Continuous delivery pipeline toolchain
- Asset and configuration management
- Production sizing and capacity planning
- Environments for integration, staging, and deployment
- Approval processes for promoting changes to next stage
- Monitoring the application and reporting errors
- SLA requirements for failover and high availability
- Restore applications following disaster
- Archiving data for auditing or support
- Upgrades to production and data migration

Operations *(SREs)* and developers should collaborate on creating and using continuous delivery practices. The following are example recommended engineering practices:

- Only build binaries once.
- Deploy the same way to every environment.

- Use idempotency—leave the target environment in the same condition as the initial state.
- Start with a known good baseline provisioned either automatically or through virtualization.
- Propagate changes through the pipeline automatically—one stage triggers the next.
- If any part of the pipeline fails in any critical way, stop the line and resolve the failure.
- Deployment Tests are designed to validate the deployment environment—for instance, by doing the following:
    - Contacting the deployment node
    - Retrieving records from deployment node
    - Asserting the message broker has registered the correct set of messages
    - Pinging through the firewall to prove that it allows traffic and provides round-robin load distribution between servers

The average time taken to **restore** a Configuration Item or IT **Service** after a Failure, *Mean-Time-to-Restore-Service (MTTRS)*, is measured from the time when the IT **Service** fails until it is fully **restored** and delivering its normal functionality. With continuous deployment, MTTRS is considered more important than *Mean-Time-Between-Failure (MTBF)* because each release has a short life before the next one is released. If a failure does occur, it is most critical to restore service quickly. The following are recommended engineering practices that help minimize *MTTRS:*

- Identify high risk changes and plan recovery actions, including a run book/event decision tree prior to release.
- Install monitors and alerts to identify the failures and notify all relevant stakeholders when an event occurs.
- Keep a chat channel open 24/7 to support communication during an event.

Given that *Continuous Delivery* and *Deployment* involves rapid changes to live production environments that customers depend on, there is always some risk that a fault will cause a customer-affecting failure. To minimize or eliminate the impacts of failures, CD employs zero downtime release strategies. *Zero Downtime* deployment strategies with gradual controlled rollouts such as *Blue-Green* and *Dark Launches* and *roll-backs* are recommended.

Blue-Green Deployments

Taking software from the final stage of testing to live production needs to do be quick to minimize downtime. The *Blue-Green* approach, illustrated in **Figure 55—Blue-Green Deployments**, use two production environments, as identical as possible as indicated in the following:

- At any time, one or the other environment is live (for this example, let's say Blue).
- Do the final release testing in the Green environment.
- Once the software is working in Green, switch the router so new requests go to Green. Blue is now idle.
- If problems occur, switch back to Blue.

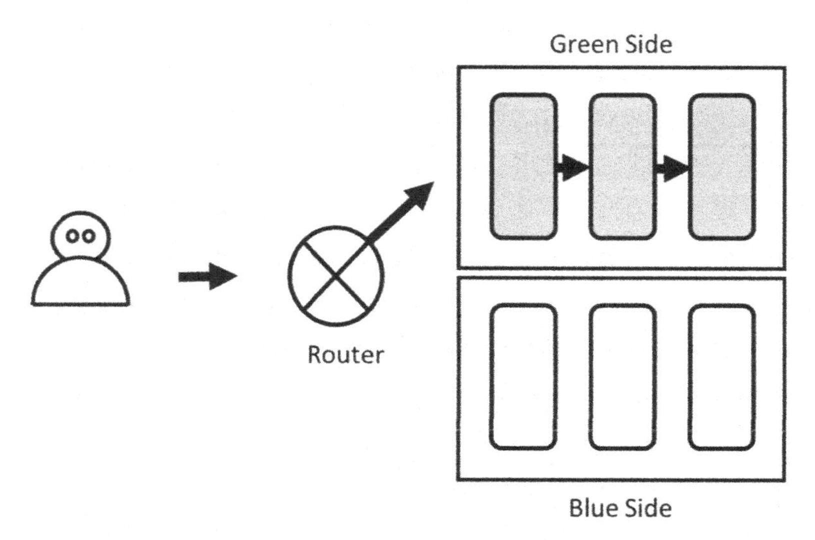

Figure 55—Blue-Green Deployments

Dark Launching Release Strategy

Dark launching is a stealthy release process that uses a combination of *Feature Toggles* and *A/B testing.*

This has the advantage of gradually releasing and testing new features to a set of users before releasing to everyone. This is a safe way to assess user acceptance and system performance.

Feature Toggles and A/B Test Strategy

The *A/B test* strategy involves a concept called *Feature-Flags*, or *Feature Toggles*. The Flags or Toggles are software switches imbedded in the executable software that can be activated through an API. Feature toggles allow changes to be selectively exposed to predetermined "trial" customers, thereby enabling controlled customer evaluations of features and alternative designs without requiring new release artifacts to be generated. **Figure 56—Feature Toggles** with A/B Testing illustrates the idea of Feature Toggles as used in *A/B testing.*

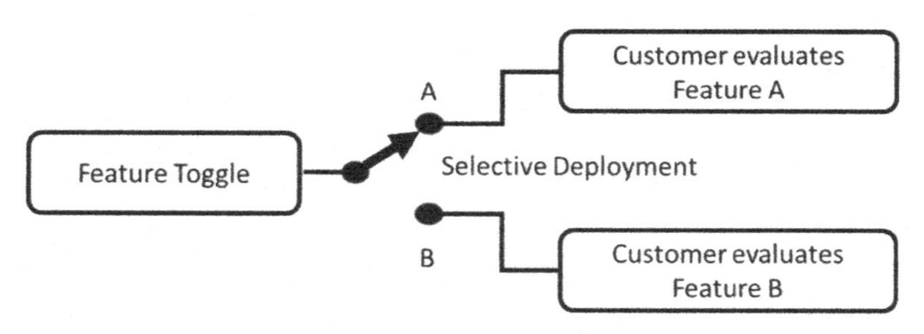

Figure 56—Feature Toggles with A/B Testing

Feature Flag Roll-Out Deployments

Feature-Flags can be used to implement a deployment strategy called a *Feature-Flag Roll-Out*, as illustrated in **Figure 57—Feature Flag Roll-Out**. *Feature-Flag Roll-Outs* recommended engineering practices are as follows:

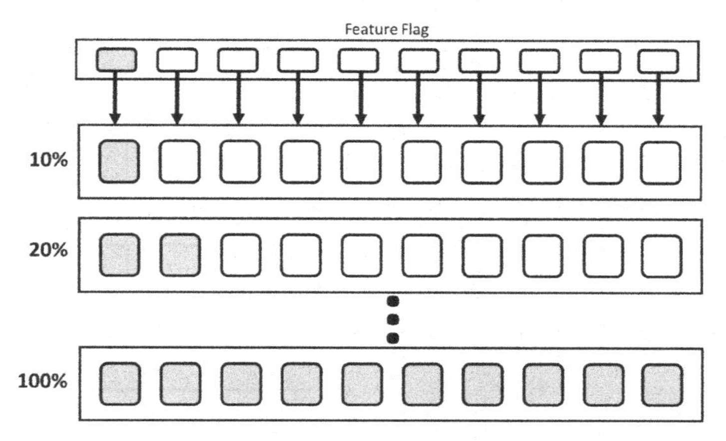

Figure 57—Feature Flag Roll-Out

- Roll out changes in parallel to a few nodes/users.
- Gradually increase to larger group.
- Direct traffic to servers to drive usage.
- Revert changes if failures are detected.
- Measure improvements using application performance monitoring tools.
- Continue the deployments until completed.

Canary Deployments

A *canary* deployment (which uses a *canary* test) is a push of code changes to a small number of end-users who have not volunteered to test anything. **Figure 58—Canary Testing** illustrates the idea. The goal is to verify code changes are transparent and work in the live production environment as intended. Like canaries that were once used in coal mining to alert miners when toxic gases reached dangerous levels, end-users selected for testing are unaware they are being used to provide an early warning of problems with the new software version. *Canary* deployments proceed with caution using a rolling deployment strategy like the *Feature-Flag* roll-out approach described in the prior section. If problems occur, the changes are rolled-back to the prior version.

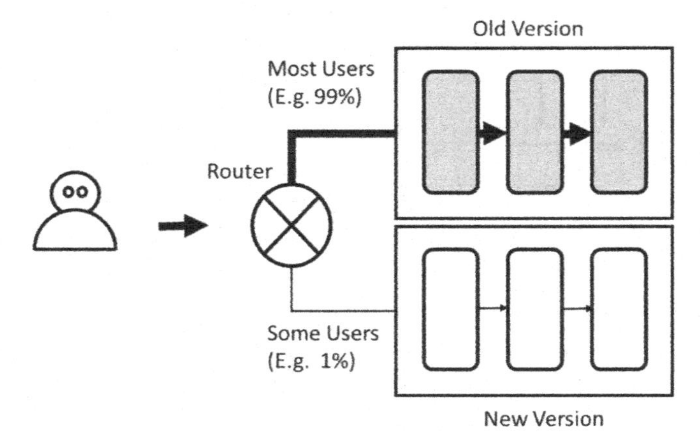

Figure 58—Canary Testing

Deploy Database Changes to Production Best Practice

Deployment of *database* changes is a special case that deserves special mention when discussing deployment strategies. The following are some recommended engineering practices to be considered when deploying database changes that need to go with application changes that are being deployed:

- Take the production database to the state required by the version of the application being deployed using the same scripts used for automated acceptance testing.
- Store delta scripts (schema + reference meta data) in a version management system and metadata table in a database to keep track of which scripts have been run against each version of the database.
- Create the new database, add the metadata table, and restore the schema and reference data to be the current production state
- Run each delta script in order. Stop if any delta fails. Record results in a metadata table.
- Acceptance test the database with the application to verify the database scripts worked.

Microservices Deployments Best Practice

Modern applications designed with a *microservice* architecture require specialized deployment strategies and tools as shown in **Figure 59—Microservices Deployment**. The following are some recommended engineering practices for *microservices* deployments:

- Deploy *microservices* independently of each other if possible. Since *microservices* are intended to be fully bounded and self-contained, this should be feasible in most cases.
- If multiple *microservices* must be deployed together, perform the rolling deployments in parallel and gradually in batches on well-bounded service clusters so it is clear where the new *microservices* are in the deployment network and can be backed out easily in case or problems.
- Spin up and destroy *microservices* on demand without disrupting service availability using immutable containers and server configurations.
- Keep staging environments as close to production as is practical to ensure tests conducted prior to deployment are valid.
- Deploy one service per host for the following reasons:
  - Minimizes the impact of one service on others
  - Minimizes the impact of a host outage

Deploying Containers with Kubernetes

Kubernetes is a portable, extensible open-source platform for managing containerized workloads and services that facilitates both declarative configuration and automation. It has a large, rapidly growing ecosystem. Kubernetes services, support, and tools are widely available.[RW41] It is not the purpose of this book to detail any tool, not even very important ones like Kubernetes that have become a de facto standard for deployment orchestration of containerized *microservices* over large-scale infrastructures, but it would be a miss not to mention some of its key capabilities and benefits.

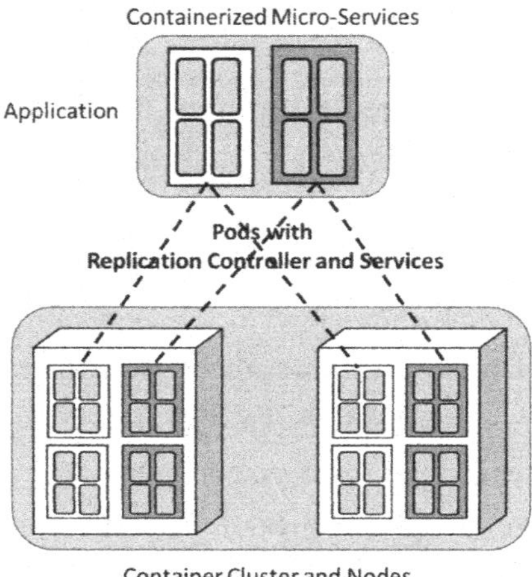

Figure 59—Microservices Deployment

Kubernetes serves several *Continuous Deployment* Use Cases, including the following:

- Creating a deployment
- Updating a deployment
- Rolling back a deployment
- Scaling a deployment
- Pausing and resuming a deployment
- Keeping track of deployment status
- Cleaning up after deployments

In Part III of this book, engineering blueprints and engineering best practices were described for Applications, Pipeline, and Infrastructures. It was shown how these three axes of DevOps can mature independently but work best when they are engineered to mature together in a balanced approach. Engineering blueprints and recommended engineering practices were presented for the Continuous Testing, Continuous Monitoring, and Continuous Delivery pillars of DevOps.

In Part IV, a seven-step DevOps transformation engineering blueprint will be shown, and engineering practices for each step will be explained.

PART IV

TRANSFORMATION ENGINEERING BLUEPRINT

"How Merlin likened the Round Table to the world, and how the knights that should achieve the Sangreal should be known. When Merlin had ordained the Round Table he said, by them which should be fellows of the Round Table the truth of the Sangreal should be well known. And men asked him how men might know them that should best do and to achieve the Sangreal? Then he said there should be three white bulls that should achieve it, and the two should be maidens, and the third should be chaste. And that one of the three should pass his father as much as the lion passeth the leopard, both of strength and hardiness.

They that heard Merlin say so said thus unto Merlin: Sithen there shall be such a knight, thou shouldest ordain by thy crafts a siege, that no man should sit in it but he all only that shall pass all other knights. Then Merlin answered that he would do so. And then he made the Siege Perilous, in the which Galahad sat in at his meat on Whitsunday last past. Now, madam, said Sir Percivale, so much have I heard of you that by my good will I will never have ado with Sir Galahad but by way of kindness; and for God's love, fair aunt, can ye teach me some way where I may find him? for much would I love the fellowship of him."

—**Le Morte d'Arthur**, *BOOK XIV CHAPTER II*

In this part, **DevOps Seven-Step Transformation Engineering Blueprint,** I provide a description and tools for the seven-step process I created for DevOps. The seven steps are presented in seven chapters: Visioning, Alignment, Assessment, Solution Engineering, Realization, Operationalize, and Expansion. By using these tools DevOps leaders and practitioners can create, implement, operate, and expand their DevOps across the organization. The chapters go further to explain how to evolve DevOps from a successful *First Way* DevOps (*Continuous Flow*) towards realizing more advanced *Second Way* (*Continuous Feedback*) and *Third Way* (*Continuous Improvement*) DevOps implementations. This part includes a discussion of **"Beyond Continuous Improvement"**—a look at emerging technologies that are shaping DevOps and how you can prepare your DevOps and yourself for the future.

This part includes a discussion of DevOps training and the importance of continuous learning.

19

DevOps Seven-Step Transformation Engineering Blueprint

To engineer is to do the noble work of engineering. DevOps engineering requires skills of elite leaders and engineers that have a profound understanding of DevOps practices that are detailed in this book. DevOps leadership and engineering requires the persistent application of skills and practices methodically toward designing solutions that achieve business and team goals. While driven by visionary ideals, DevOps engineering requires practical, disciplined, progressively refined implementations using carefully chosen dimensions of people, process, and technology solutions. At any point in the DevOps engineering life cycle, the goal is to achieve a solution that balances the Nine Pillars of DevOps for a consistent maturity level while questing the Holy Grail—*Continuous Improvement.*

Figure 60—DevOps Seven-Step Transformation Engineering Blueprint prescribes an infinite cycle of seven steps for achieving your DevOps goals methodically, no matter what your goals or level of DevOps maturity are currently.

Transforming from one maturity level of DevOps to the next higher level requires transitioning through the seven steps of each improvement cycle to effect progressive refinements necessary for a solution that maintains a balance across the *Three Dimensions of DevOps* (People, Process, and Technology) and *The Nine Pillars of DevOps*—Leadership, Collaborative Culture, Design for DevOps, Continuous Integration, Elastic Infrastructure, Continuous Monitoring, Continuous Security, and Continuous Delivery/

Deployment. When the *Dimensions* and *Pillars* are out of balance, *Continuous Flow* is difficult to maintain and goals will be impacted. Changes to the *Dimensions* and *Pillars* are introduced incrementally and tested methodically for each cycle in a continuous improvement loop.

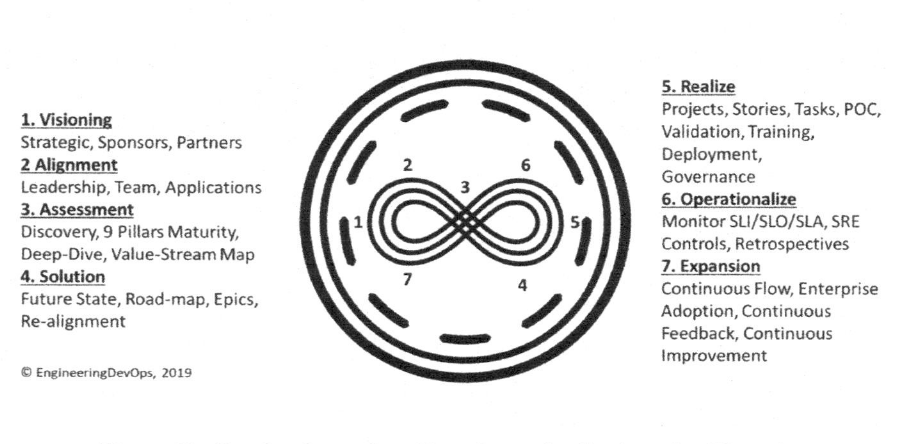

Figure 60—DevOps Seven-Step Transformation Engineering Blueprint

The seven steps of the DevOps Transformation Engineering Blueprint are detailed as follows:

1. **Visioning**—Define the strategic need for DevOps for organization and identify sponsors that will own the DevOps transformation at the strategic level, as well as key partner organizations that need to be strategically aligned to the DevOps transformation.
2. **Alignment**—Leaders and key team members that are most important to each DevOps transformation cycle agree to specific goals for selected applications.
3. **Assessment**—For the current state of selected applications, DevOps maturity is discovered and assessed, deep-dive *assessments* are conducted for specific topics, and a current state value-stream map is created relative to the organization goals.
4. **Solution**—A DevOps expert consultant performs analysis of assessment data and formulates a future state *value stream* road-

map including *Themes*, *Epics*, and *User Stories* and obtains alignment with key stakeholders.

5. **Realize**—DevOps implementation projects are defined including user stories and tasks, product *Proof of Concept (POC)* trials are conducted to validate solution choices, the solution is validated with selected applications and use cases, training is conducted as the solution is deployed to production, and governance practices for the new solution are activated.
6. **Operationalize**—Deployed DevOps improvements are monitored and controlled with *Site Reliability Engineering (SRE)* practices that monitor *SLI*, *SLO*, and *SLA* metrics. Retrospectives are conducted to create actionable prioritized lessons learned for continuous improvement.
7. **Expansion**—Once *Continuous Flow* (*The First Way* of DevOps) is realized for a select set of applications, the organization can safely expand the solutions to other applications across the organization. Further transformation cycles will lead to realization of *Continuous Feedback* (*The Second Way* of DevOps) and *Continuous Improvement* (*The Third Way* of DevOps).

Recommended engineering practices and tools to implement the seven steps of DevOps Transformation Engineering Blueprint are explained in the following chapters.

20

Step One: Visioning

In this step, the sponsors and strategic leaders that will own the DevOps transformation from one level of DevOps maturity to the next are identified, key partner organizations that need to be strategically aligned to the DevOps transformation are identified, strategic goals for DevOps for the organization are defined, and actions towards the next step are committed.

Why Is the Visioning step important to DevOps Engineering?

DevOps transformations take considerable amount of time and resources. Short-term tactical DevOps projects may not survive long enough or enjoy sufficient support to realize changes needed to meet strategic DevOps goals.

If strategic-level leadership and essential partners do not have clear agreement that DevOps is a strategic priority, then the transformation projects will likely not get sufficient investment of resources or have the collaborative support needed to be successful.

How Is the Visioning Step Accomplished?

A DevOps Transformation Initiator (*Initiator*) decides that DevOps is a priority for the organization. The Initiator may be practically anybody, but usually it is a manager or influencer with a passion for DevOps derived from an understanding of the values that DevOps can bring to the organization.

At least one DevOps Transformation Sponsor (*Sponsor*) for the DevOps transformation is identified. This may be a qualified *Initiator*, or it may be someone else. The *Sponsor* needs to be a recognised change agent with direct authority or strong influence over leaders and key partners for all the functions that are most relevant to the DevOps transformation. The *Sponsor* needs to be committed to sponsoring the DevOps transformation, including raising funding and defining and defending DevOps program priorities.

A DevOps Transformation Expert (*Expert*) is identified. This may be a third-party consultant or someone within the organization that has a deep understanding and experience with DevOps at the target maturity level. The role of the Expert is to provide technical advice and "hands-on" continuity through the seven transformation steps.

A DevOps Transformation Visioning Meeting is conducted with the *Initiator*, *Sponsor*, and *Expert* to define the strategic goals, approach, key players, and next steps. An example of the DevOps Transformation Visioning Meeting result is provided in Appendix C. The following is covered in the meeting:

1. Use the DevOps Transformation Application Scorecard provided in Appendix B to select the application or set of applications that will be targeted for the DevOps Transformation and identify a model application to start the Transformation. Organizations may have multiple applications that are transforming to a DevOps approach for improving development and operations. Transforming all of them at once is generally exceedingly expensive, difficult to coordinate, and not a prescription for successful DevOps adoption at the enterprise level. The engineering solution is to choose one application that can be used as a model for DevOps transformations. Some of the key reasons and benefits picking a model application for enterprise DevOps transformations are listed as follows:

a. DevOps experimentation, implementation, and improvements are visible to the enterprise.
b. Demonstrate ROI for DevOps projects.
c. Provide a Reference Implementation that other application teams can learn from and apply to their applications

2. Define and document strategic goals for the DevOps transformation. At this level, the goals are expressed as visionary qualifiable but not necessarily quantifiable objectives, which will have a major impact on the purpose of the organization. These are not technical goals as much as they are business goals. For example, statements about improvements to competitiveness, innovation, customer satisfaction, and employee satisfaction would qualify as visionary strategic goals. Technologies expected to accomplish these goals may be included, but the emphasis of these goals are on strategic outcomes instead of the means to accomplish them. It is important to explain why this goal was chosen over alternatives considered.
3. Agree to follow the DevOps Transformation Engineering Blueprint.
4. Identify strategic level leaders and partners for the DevOps Transformation Team that cover the cross-section of organization functions that are most relevant to the DevOps transformation. For example, strategic leaders with responsibility over roles from Development, Quality Assurance, Operations, Infrastructure, DevOps Tools, Third-Party Suppliers, Product Owners, Security, Project Management, Training, Finance, Human Resources, and Governance should be considered, depending on the extent to which each of the roles are likely to be affected by the transformation.
5. Commit the following actions towards the *Alignment* step:
    - Engage a DevOps *Expert.*
    - Communicate membership, strategic goals, and kick-off meeting schedule to the DevOps *Transformation Team.*
    - Prepare the DevOps Transformation *Alignment* kick-off meeting.

Overcoming Challenges with the Visioning Step

If the *Initiator* does not qualify as a *Sponsor* or *Expert* and cannot find a qualified *Sponsor* or *Expert*, then the *Initiator* may complete this step without a *Sponsor* or *Expert*, with the intention that a *Sponsor* and *Expert* will be identified once more details of the initiative are clarified. A word of caution—starting the next step (*Alignment*) without a *Sponsor* or *Expert* is not recommended, because getting agreement amongst these three roles will accelerate the remaining steps.

If it is unclear which strategic level leaders will be needed for the specific DevOps transformation, then is it best to include leaders that are potentially important.

If the team is having trouble selecting the DevOps Transformation Application, then it is recommended that leaders consult with their teams and fill out the checklist together.

21

Step Two: Alignment

After the Vision step is completed, it is important for those leaders to get their teams aligned around the vision. In the *Alignment* step, the DevOps Transformation Team, consisting of leaders and key team members that are most important to each DevOps transformation cycle, meet to agree to on specific goals for selected applications.

Why Is the Alignment Step Important to Engineering DevOps?

- Nearly all DevOps transformation projects require cooperation and collaboration between a cross-functional team.
- Cooperatively creating a commonly agreed-upon and documented DevOps goal priority that is specific enough to measure will clarify collaboration actions for the rest of the DevOps Transformation project.
- The DevOps *Transformation Goal* is measurable and will provide the means to validate progress and judge completion of DevOps Transformation Projects.
- Proceeding to subsequent steps in the DevOps Transformation without alignment across the DevOps Transformation Team is a prescription for failure.

How Is the Alignment Step Accomplished?

The DevOps Transformation Initiator, with visible endorsement from the DevOps Transformation Sponsor, conducts a DevOps Transformation Alignment Meeting with the DevOps Transformation Team. An example of the DevOps Transformation Alignment Meeting is provided in Appendix E.

In this meeting, the following occurs:

- The *DevOps Strategic Goal* is communicated by the Sponsor
- The *DevOps Transformation Team* and roles are introduced.
- A *Definition of DevOps* is discussed and agreed.
- The *DevOps Engineering Blueprint* is presented.
- The Chosen applications for the Transformation are presented together with the scorecard.
- The DevOps Transformation Engineering Blueprint is presented to ensure everyone is clear about the big picture steps to accomplish the transformation.
- The DevOps *Expert* conducts a DevOps Transformation Goal Workshop. During this workshop, the current state and desired future state is discussed for specific goals organized in six DevOps goal categories: Agility, Stability, Efficiency, Quality, Security, and Satisfaction. For each goal, an importance score, current state, and desired future state and target date are discussed, and a consensus view is documented. The highest-priority *DevOps Transformation Goals* are determined from the workshop and are documented. These goals are important because they are used to drive priorities and activities during the remaining steps of the DevOps Transformation. The DevOps Transformation Goal Scorecard template is provided in Appendix D. An example is shown in **Figure 61—DevOps Transformation Goal Scorecard Example**.
- The team decides key recommended engineering practices areas to focus on during the assessment step using a DevOps Transformation Practices Topics Scorecard. The DevOps Transformation Practices Topics Scorecard template is provided in Appendix F. An

example is show in **Figure 62—DevOps Transformation Practices Topics Scorecard Example.**

- *Assessment* input requirements for the next step are discussed.
- Next steps actions towards the *Alignment* step are committed, including the following:
  - Schedule date to complete Alignment Data
  - Schedule date to complete Assessment inputs
  - Set Date for DevOps Transformation Assessment Workshops
  - Logistics for DevOps Transformation Assessment Workshops

Overcoming Challenges with the Alignment Step

The main challenge with the Alignment step is getting the attention of such a large group of leaders in a dedicated fashion for the meeting. To overcome this obstacle, it is critical for the *Sponsor* to encourage attendance. The meeting should be completed no more than a half-day so the participants will have time the same day to take care of other business.

Some of the current state data may not be available in the meeting. Document the best estimate, take an action to get the actual date, and move on.

It may be difficult to get consensus. The *Sponsor*, *Initiator*, and *Expert* may have to make a judgement call on which answer to accept to avoid stalling progress. If there are significant objections, then document them and take an action to discuss offline.

DevOps Transformation Goals Scorecard	Metric Unit	Importance (I) (1-5) (1=low, 5=critical)	Current State	Desired State	Percent Improvement (P%)	SCORE I x P%	RANK
Agility		4.4			205%	9	1
Lead time: Duration from code commit until code is ready to be deployed to production.	# days	5	5.0	2.0	250%	13	2
Release Cadence: Frequency of having releases ready for deployment to live production.	# releases / month	5	0.2	0.6	300%	15	1
Fraction of Non-Value Added-Time Fraction of their time employees are not spending on new value enhancing work such as new features or code.	Fraction %	3	30%	20%	150%	5	15
Batch size: Product teams break work into small batch increments.	1-10 (1 rarely, 10 usually)	5	3.0	5.0	167%	8	3
Visible Work: Workflow is visible throughout the pipeline.	1-10 (1 rarely, 10 usually)	4	5.0	8.0	160%	6	6
Security		2.7			233%	6	2
Security Events: # times that a serious business impacting security event occur over a set period	# per year	2	2	1	400%	8	4
Unauthorized Access: # times per period that unauthorized users accessed unathorized information.	# per year	3	1	1	100%	3	21
Fraction of time remediating security problems: Average % of time that employees spend remediating security issues.	%	3	5%	3%	200%	6	9
Satisfaction		3.8			137%	5	5
Employee satisfaction with team: Employees are likely to recommend their team as a great to work with.	1-10 (1 not likely, 10 most likely)	3	7.0	8.0	114%	3	20
Employee satisfaction with organization: Employees are likey to recommend their organization as a organization to work in.	1-10 (1 rarely, 10 most likely)	4	6.0	8.0	133%	5	13
Organization type: The culture is of the organization is a ***generative*** type, with good communication flow, cooperation and trustful.	1-10 (1 rarely, 10 very much)	4	5.0	7.0	140%	6	11
Leader Style for Recognition: Leaders promote personal recognition by commending team for better-than-average work, acknowledging improvement in quality of work and personally compliments individuals' outstanding work.	1-10 (1 rarely, 10 very much)	4	5.0	8.0	160%	6	6
Stability		4.0			152%	6	3
MTTR: Mean-Time-To-Recover (MTTR) from failure/service outage in production.	hours	4	1.0	0.7	154%	6	8
Code merges problems: % of code merges from development branches to the trunk branch break the trunk branch.	%	4	15%	10%	150%	6	10
Quality		3.7			148%	5	4
Failures in Production: Frequently of failures requiring immediate remediation occur in live production.	# per week	4	0.1	0.05	200%	8	4
Test and Data Available: Tests and test data are sufficient and readily available when needed.	1-10 (1 rarely, 10 usually)	3	7.0	9.0	129%	4	19
Customer Feedback: The organization regularily seeks customer feedback and incorporates the feedback into design.	1-10 (1 rarely, 10 usually)	4	7.0	8.0	114%	5	14
Efficiency		3.3			143%	5	6
Unplanned work: % of time do employees spend on all types of unplanned work, including rework.	%	3	15%	10%	150%	5	15
Operating Costs are Visible: Comprehensive metrics are kept for operating costs of development and operations.	1-10 (1 rarely, 10 usually)	3	5.0	7.0	140%	4	17
Capital costs are visable: Comprehensive metrics are kept for capital costs of development and operations.	1-10 (1 rarely, 10 usually)	3	5.0	7.0	140%	4	17
Backlog Visibility: Lean product management is practiced using highly visible, easy-to-understand presentation formats that show work to be done.	1-10 (1 rarely, 10 usually)	4	5.0	7.0	140%	6	11

The DevOps Transformation Goal Scorecard spreadsheet is posted on www.EngineeirngDevOps.com

Figure 61—DevOps Transformation Goal Scorecard Example

DevOps Transformation Practices Topics

Nine Pillars of DevOps Practice Topics	Importance (I) (1-5) (1=low, 5=critical)	Current Level of Practice (P) (1-5) (1=not yet, 5=always)	SCORE (1-5) 6-(((Ix(5-P))/5)+1)	Rank
Collaborative Leadership Practices	2	3	4.20	7
Collaborative Culture Practices	4	3	3.40	4
Design for DevOps Practices	5	2	2.00	2
Continuous Integration Practices	3	3	3.80	4
Continuous Testing Practices	3	4	4.40	7
Elastic Infrastructures Practices	3	4	4.40	7
Continuous Monitoring Practices	5	2	2.00	2
Continuous Security Practices	4	1	1.80	1
Continuous Delivery/Deployment Practices	3	3	3.80	7

Special DevOps Deep Dive Practice Topics	Importance (I) (1-5) (1=low, 5=critical)	Current Level of Practice (P) (1-5) (1=not yet, 5=always)	SCORE (1-5) 6-(((Ix(5-P))/5)+1)	Rank
DevOps Version Management Practices	5	4	4.00	11
Value Stream Management DevOps Practices	4	2	2.60	2
Application Release Automation Practices	4	3	3.40	8
DevOps Infrastructure-As-Code	3	2	3.20	4
Hybrid Cloud DevOps Practices	3	2	3.20	4
Multi-Cloud DevOps Practices	3	2	3.20	4
Application Performance Monitoring Practices	4	3	3.40	8
DevOps Training Practices	3	2	3.20	4
Site Reliability Engineering Practices	3	1	2.60	2
DevOps Service Catalog	3	3	3.80	10
DevOps Governance Practices	4	1	1.80	1
Other >	0	0	5.00	12

The DevOps Transformation Practices Topics Scorecard is posted in spreadsheet form on www.EngineeringDevOps.com

Figure 62—DevOps Transformation Practices Topics Scorecard Example

22

Step Three: Assessment

After the key leaders and team members are aligned around DevOps goals by completing Step Two, the Assessment step can proceed. In the *Assessment* step of the *DevOps Transformation Engineering Blueprint*, the current state of selected Applications is discovered, maturity of DevOps practices *assessments* and Deep-Dive practices assessments for specific topics are conducted, a current state *value-stream map* is created relative to the organization goals, and alignment around solution requirements is obtained with stakeholders.

Why Is the Assessment Step Important to Engineering DevOps?

Assessments embody the data, expertise, and alignments necessary to engineer a DevOps solution that realizes the DevOps goals. Without assessments, the DevOps Transformation solution is an ad-hoc effort instead of an engineered solution and will not accomplish or sustain goals.

How Is the Assessment Step Accomplished?

The following are the steps of a DevOps Assessment:

- Discover current state
- Assess the maturity of DevOps Practices
- Create a Current State *Value-Stream Map*
- Align priorities for solution requirements

Figure 63—DevOps Assessment Engineering Blueprint illustrates that inputs from the prior Alignment step prescribe a DevOps definition, a selected Application for *Assessment*, *DevOps Goals*, and *Assessment* priorities. The surveys and workshops that are part of the *Assessment* step are customized in accordance with these prescriptions.

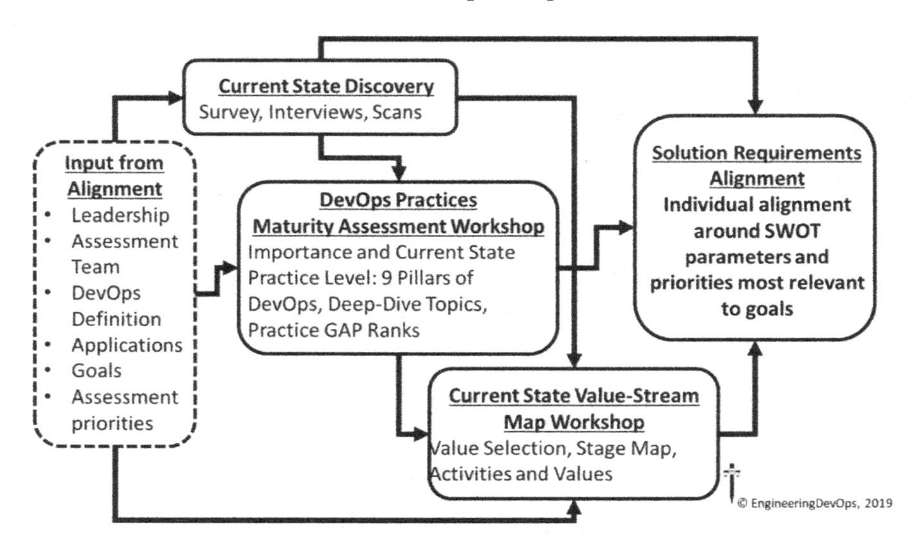

Figure 63—DevOps Assessment Engineering Blueprint

Discover Current State

The *DevOps Assessment Discovery Survey* presented in Appendix G is sent to staff members to gather current state information for the Application, Organization, Pipeline, Tools, Infrastructure, and COTS systems that are relevant for the Application. This survey provides information for the Assessment and Value-Stream workshops and Solution Requirements Alignment.

Assess the Maturity of DevOps Practices

A DevOps Practices Maturity Assessment Workshop is held with the DevOps Transformation team to assign scores to *The Nine Pillars of DevOps* and other Deep-Dive topics that were selected during the Vision step. Appendix H shows the format for the workshop. The output of this workshop identifies which practices are ranked the highest and is used as input to the Current State *Value-Stream Mapping* Workshop and the *Solution Requirements* Alignment step of Assessment. The DevOps Practices Maturity Workshop uses an Assessment tool to help gather, score, and rank the practices. An example of the tool is shown in **Figure 64—DevOps Practices Maturity Assessment Tool Example**.

	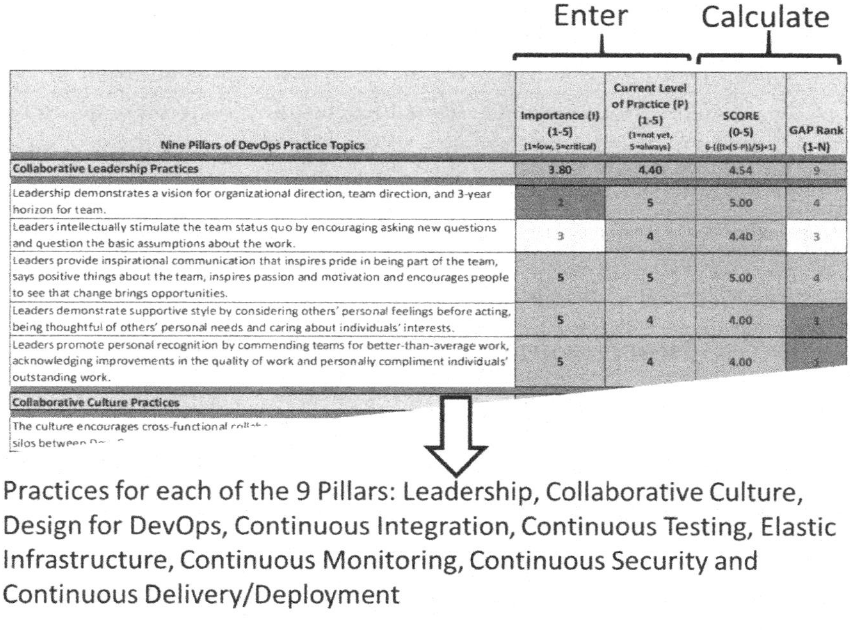 Enter		Calculate	
Nine Pillars of DevOps Practice Topics	Importance (I) (1-5) (1=low, 5=critical)	Current Level of Practice (P) (1-5) (1=not yet, 5=always)	SCORE (0-5) 6-(((Ix(5-P))/5)+1)	GAP Rank (1-N)
Collaborative Leadership Practices	3.80	4.40	4.54	9
Leadership demonstrates a vision for organizational direction, team direction, and 3-year horizon for team.	2	5	5.00	4
Leaders intellectually stimulate the team status quo by encouraging asking new questions and question the basic assumptions about the work.	3	4	4.40	3
Leaders provide inspirational communication that inspires pride in being part of the team, says positive things about the team, inspires passion and motivation and encourages people to see that change brings opportunities.	5	5	5.00	4
Leaders demonstrate supportive style by considering others' personal feelings before acting, being thoughtful of others' personal needs and caring about individuals' interests.	5	4	4.00	1
Leaders promote personal recognition by commending teams for better-than-average work, acknowledging improvements in the quality of work and personally compliment individuals' outstanding work.	5	4	4.00	1
Collaborative Culture Practices				
The culture encourages cross-functional ... silos between ...				

Practices for each of the 9 Pillars: Leadership, Collaborative Culture, Design for DevOps, Continuous Integration, Continuous Testing, Elastic Infrastructure, Continuous Monitoring, Continuous Security and Continuous Delivery/Deployment

Figure 64—DevOps Practices Maturity Assessment Tool Example

Create a Current State Value-Stream Map

A Current State *Value-Stream Mapping* Workshop is held with the *DevOps Transformation* team to identify pipeline stages, activities, and values that are most relevant to accomplishing the DevOps goals. Appendix I shows the format for the workshop. The output of the workshop includes bottlenecks and items that may not have been identified in prior steps. Any new items are added to a list for review during Solution Requirements Alignment. A template for the *Value-Stream Map* is provided in Appendix K. An example of a current state value-stream map is shown in **Figure 11—Value-Stream Map Example**.

Align Priorities for Solution Requirements

As shown in **Figure 65—DevOps Engineering Solution Requirements Matrix** Solution, Requirements Alignment involves assembling all the highest-ranked practices and items identified in the DevOps Practices Maturity Assessment and Current State Value-Stream Map workshops. The highest priority items in the list, after sorting inputs, are used to survey the *DevOps Transformation Team* to determine their view of overall priorities. The highest priority outputs are the ones that get the greatest number of votes and become the solution requirements. Since this is the result of progressive team inputs and consolidated team votes, the solution requirements are well-aligned to the *DevOps Transformation Team's* preferences.

Overcoming Challenges with the Assessment Step

High-quality discovery survey data and full participation in the workshops is vital to the success of the *Assessment* step. If the DevOps *Expert* determines that the information and participation are not sufficient, it may be necessary to halt the *Assessment* process, and the Sponsor may have to press for action before resuming.

If the DevOps *Expert* determines that there are too many high-priority solution requirements to practically solve in one DevOps transformation cycle, then the requirements need to be portioned into phases of requirements. One way to do this is to take a second pass vote on the priority list by only including the highest-ranked items resulting from the prior vote.

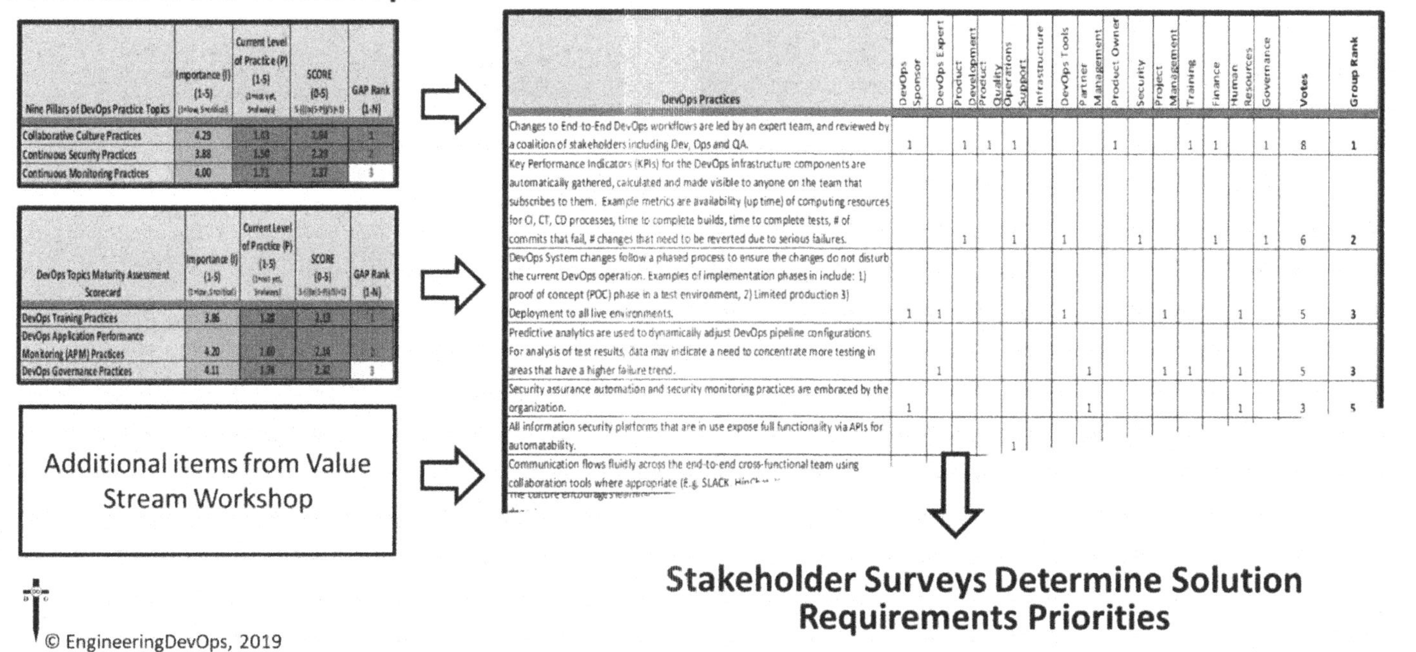

Figure 65—DevOps Engineering Solution Requirements Matrix

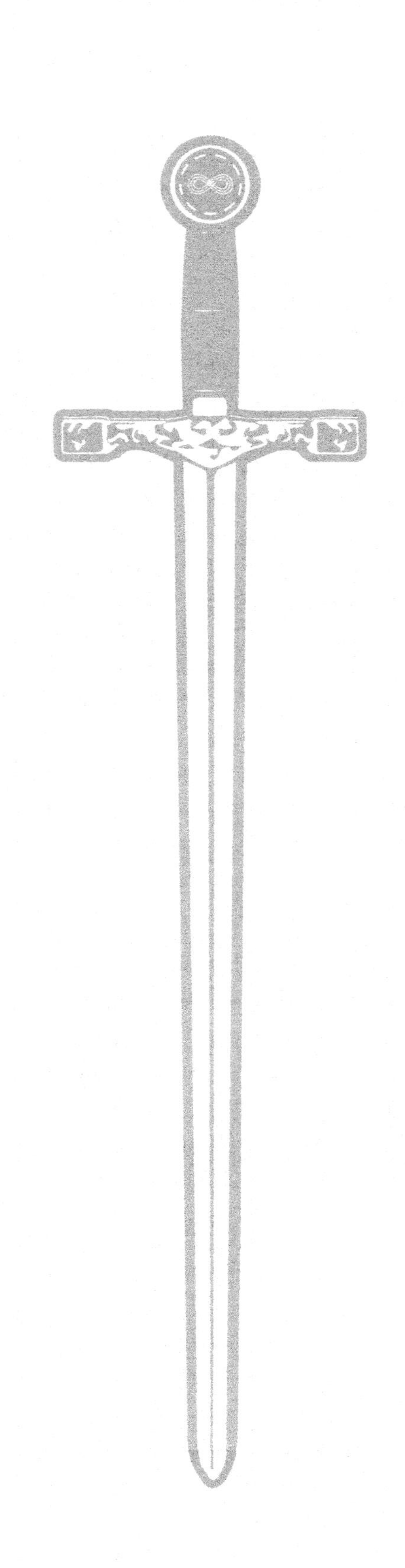

23

Step Four: Solution

In the Solution step, the DevOps *Expert* performs analysis of DevOps Solution Requirements that was developed in Step Three. *The DevOps Expert* formulates a future state *Value-Stream Map* and *Solution Roadmap* including Themes, Epics, and User Stories to obtain solution alignment with the *DevOps Transformation Team.*

Why Is the Solution Step Important to Engineering DevOps?

The engineering methodology described in this book, if followed carefully, will result in a successful, if not unique or deterministic, DevOps transformation. Nevertheless, the solution that results from following the methodology in this book is not the only possibility. There are many possible viable solutions to DevOps Transformations. It would be great if there was a methodology that always yields a recommended solution that is most optimal, but this is not practical given the large number of components and variables involved with *The Three Dimensions*, *Nine Pillars*, and *Twenty-Seven Critical Success Factors* of DevOps Transformations. However, for every viable solution, there are more solution choices that will lead to failure.

A failed solution can have disastrous consequences, such as the following:

- Lost business opportunity costs caused from delays
- Poor ROI due to overrun labor and systems costs
- Morale problems

Following the methodology in this book with detailed and persistent engineering discipline will yield a solution that will meet the DevOps goals.

How Is the Solution Step Accomplished?

The *DevOps Transformation Solution* step includes the following substeps and is illustrated in F**igure 66—DevOps Transformation Solution Engineering Blueprint**:

- Create a Future State *Value-Stream Map*
- Determine tools recommendations
- Road-map the DevOps Transformation
- Build a backlog of themes, epics, and user stories
- Estimate ROI
- Solution recommendation alignment

Create a Future State Value-Stream Map

A Future State *Value-Stream Map* is created by the DevOps Expert, given the current stated *Value-Stream Map*, *DevOps Transformation Solution Requirements*, and goals. The new value stream identifies changes to the DevOps stages and *The Three Dimensions of DevOps* (People, Process, and Technology) needed to accomplish the goals. The output of the workshop includes identifying specific targets to reduce bottlenecks for have been identified in prior steps. A template for the *Value-Stream Map* is provided in Appendix K.

DevOps Tools Recommendations

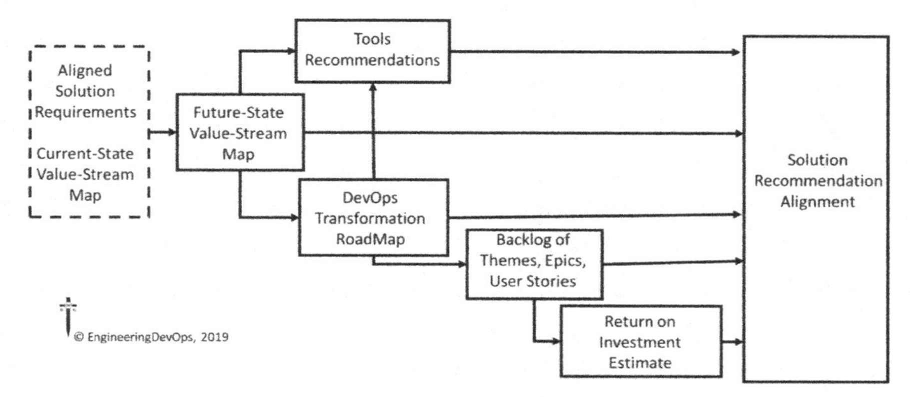

Figure 66—DevOps Transformation Solution Engineering Blueprint

The Future-State *Value-Stream Map* determine the types of tools that are needed for the DevOps Transformation. Appendix L provides a chart with some of the most popular tools used with DevOps at the time of this writing. DevOps tools is a highly competitive marketplace, with tool choices changing frequently. There are many sources for tools information such as *The DevOps State of DevOps Tools* report[RW42] and the XebiaLabs Periodic Table of Tools chart.[RW43] Gartner[RR9] and Forrester[RR10] are good sources for comparison information. Tools vendors generally have comparison charts for other tools they compete with.

There are generally the following four major sources for tools:

- ***Open-Source:*** Unlimited distribution without cost with support and evolution provided by the open-source community
- ***Freemium:*** Limited distribution with minimal or no cost with support and evolution provided by a commercial vendor
- ***Do-It-Yourself (DIY)*** or ***"Homegrown":*** Tools solutions with unlimited distribution within the source organization with support and evolution provided by the source company
- ***Enterprise License:*** Limited distribution under commercial license with support and evolution provided by a commercial vendor

The selection of tools for Engineering a DevOps transformation considers the following **DevOps tool requirements** factors:

- **Cost:** The licensing and labor cost for an initial configuration and total cost of ownership at scale, including costs internal to the organization
- **Compatibility:** Operating systems, ecosystems, cloud-native, DevOps frameworks, APIs
- **Ease of Use:** Intuitive user interface, efficient controls, efficient outputs
- **Administration Capabilities:** Installation support, diagnostic support, built-in metrics to track performance
- **Functional Requirements:** Features and capabilities
- **Non-Functional Requirements:** Performance, reliability, stability, scalability
- **Roadmap:** Planned enhancements for future solution requirements
- **Support:** Professional services for installation, proof-of-concept, configuration, training, and bug fixes

Preparing a DevOps tools comparison matrix, such as the one shown in Appendix L, that lists the DevOps tool requirements side by side with the characteristics of alternative tools is the first step to decide which tool or smaller set of tools best fits a specific DevOps transformation. If there are multiple tools that are equally well suited, then a deeper dive or *Proof-of-Concept (POC)* may be required during the Realize step of the *Engineering DevOps Transformation.*

Road-Mapping DevOps Transformation

The Future-State *Value-Stream Map*, together with the *DevOps Transformation Solution* Requirements, determine the major Themes, Epics, and User Stories required to satisfy the DevOps Transformation goals. **Figure 67—Engineering DevOps Transformation RoadMap Example** shows how each *Theme* is represented as a block of *Epics*, each of which have five

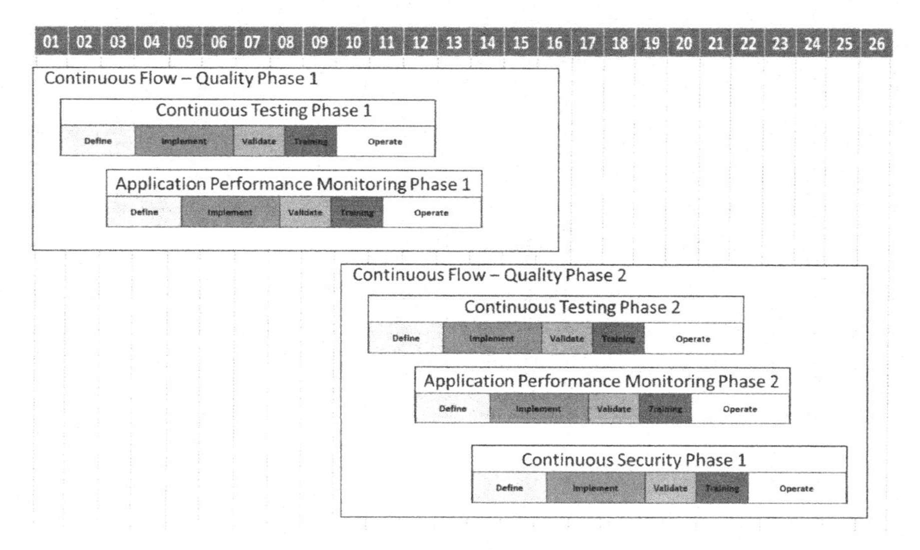

Figure 67—Engineering DevOps Transformation RoadMap Example

EPIC	Phase	User Story	Responsible
Continuous Testing Phase 1	Define	As a developer I want to use Test Driven Development to define my unit tests before I write my code.	Developer
Continuous Testing Phase 1	Implement	As a developer I want to use Test Driven Development to define my unit tests before I write my code.	Project Manager
Continuous Testing Phase 1	Validation	As a developer I want to use Test Driven Development to define my unit tests before I write my code.	QA Engineer
Continuous Testing Phase 1	Training	As a developer I want to use Test Driven Development to define my unit tests before I write my code.	QA Engineer
Continuous Testing Phase 1	Operate	As a developer I want to use Test Driven Development to define my unit tests before I write my code.	Developer

Figure 68—DevOps Backlog Example

phases: Definition, Implementation, Validate, Training, and Operation. Estimates for each Epic time duration can be determined by estimating the average time for implementing a user story by an implementation team and dividing by the number of implementation teams available to implement them.

Figure 68—DevOps Backlog Example shows a backlog for one Theme-Epic-User Story combination. It is important to identify all the User Stories and responsible owners for them to make sure the owners agree before proceeding to the next steps in the Engineering DevOps Transformation Solution step.

Estimate ROI

Business managers will not likely approve a solution unless it will yield a ROI. In this step, the DevOps Expert, together with the Responsible Owners of the User Stories for each Epic and someone from the Finance department, produce an estimated ROI for each Epic. Appendix O presents a DevOps ROI Calculator. **Figure 69—DevOps ROI Calculation Example** a ROI calculation for an example Epic.

There are several key elements in the DevOps *ROI* calculation worth noting, detailed as follows:

- Involve someone from the Finance department. It is important that they agree to the *ROI* data, or the solution proposal may be rejected on their recommendation alone.
- The DevOps ROI needs to be sufficiently compelling compared to alternative projects that the organization is considering in its investment portfolio. It is important to know the ROI thresholds for the organization's investments. The DevOps ROI must be above the investment threshold to qualify for investment.
- Keep release costs with and without the DevOps enhancements separated from the DevOps enhancements implementation costs.
- A multiyear view, typically the same as the capital depreciation schedule used by the organization, is recommended. The initial year costs of implementing a DevOps enhancement generally does not yield sufficient ROI to justify the cost. DevOps enhancements generally yield the largest cost benefits for years following the initial implementation year.
- Be careful to compare "apples to apples" when comparing costs between the "No-DevOps enhancements" and the "DevOps

Per Application (3 years Period)	Formula	NO DevOps Enhancements (n)	DevOps Enhancements (d)
R) Average # releases/year	**Estimate Number**	6	10
A) Total Costs (3 years) $K	B+C	**$7,550**	**$6,480**
B) Cost for 3 years of releases $K	**B1 + B4**	**$7,550**	**$5,220**
B1) Labor Costs $K	3 x B2 x B3	$6,600	$4,620
B2) Average Labor rate $K/year	Estimate Number	$110	$110
B3) Average # workers	Estimate Number	20	14
B4) Capital Depreciation (3 years amortization) $K	Estimate Number	$950	$600
C) DevOps Enhancements costs $K	**C1 + C4**	**$0**	**$1,260**
C1) Labor Costs$K	3 x C2 x C3	$0	$660
C2) Average Labor rate $/year	Estimate Number	$110	$110
C3) # workers for DevOps enhancement	Estimate Number	0	2
C4) Capital Depreciation (3 years amortization) $K	Estimate Number	$0	$600
D) Direct Savings Attributed to DevOps Enhancements	**An-Ad**		**$1,070**
F) Cost per release	**A/R**	**$419**	**$216**
H) # Releases over 3 years	3 x R	18	30
I) Additional releases due to DevOps Enhancements	Hd - Hn		12
J) Costs of Equivalent # releases with DevOps enhancements (3 years) $K	F x Hd	$12,583	$6,480
K) Equivalent Savings due to DevOps Enhancements (3 years) $K	Jd - Jn		$6,103
L) Return on Investment (ROI)	K / C		5
M) Payback period (Months)	C / (K/36)		7
E) Number of Applications	Estimate Number	5	5
O) Direct Savings for all applications (3 years) $K	E x Dd		5,350
P) Equivalent Savings for all applications (3 years) $K	E x Kd		30,517

Figure 69—DevOps ROI Calculation Example

Enhancements" alternatives. The ROI example does this by only comparing the same number of releases between the two columns.

- Remember to include not only the initial Application but also all Applications for which the DevOps enhancement will apply over the comparison period.

Solution Recommendation Alignment

A DevOps Solution Recommendation Meeting is held with all stakeholders that need to agree to the DevOps Transformation. This includes DevOps *Sponsor*, *Initiator*, *Expert*, and *DevOps Transformation Team* Leaders together with the *Business Leaders*.

During this meeting, the following topics are discussed:

- DevOps Transformation Goals
- Summary of Current State of DevOps
- Summary of DevOps Solution Requirements
- Future-State Value-Stream Map
- DevOps Transformation RoadMap
- ROI
- Solution Recommendation Summary
- Next Steps

A template for the Solution Recommendation Alignment Meeting is provided in Appendix P.

Overcoming Challenges with the Solution Step

Estimates required for completing the Future-State *Value-Stream Map* and the *ROI* calculations for *Epics* rely on experience of the DevOps *Expert* and *DevOps Transformation Team* members. To ensure the numbers are believed by *Sponsor* and *Business Leaders* the *Expert* and *DevOps Transformation Team* members must be clearly in alignment and be prepared to provide details and sources to back up the estimates.

Tools choices can get bogged down in "religious preferences." It is important to have an objective comparison matrix to support tool recommendations. Tool comparisons based on strategic factors are more likely to be accepted.

ROI values for DevOps projects must meet thresholds for the organization's investments over an equivalent ROI period. A short payback period is important to emphasize; however, be careful not to fall into the trap and promise that "it won't cost anything because it will pay for itself this year." The business leaders may then accept that you can do the project without any investment in that case and ask to see the reduction this year!

24

Step Five: Realize

Realizing DevOps implementations is mostly the same as Application software development, right? After all, coding is coding. Wrong! There are similarities but also critical differences that need to be considered during the Realize step of the DevOps Transformation.

Certainly, similarities do exist. Both DevOps projects and Application development projects involve designing, coding, integration, testing, validation, deployment, and operation of code changes. Many of the same processes and tools can be shared between developers of DevOps changes and developers of applications.

Differences, however, are critical to fully understand and consider during the Realize step when engineering a *DevOps Transformation*. The following are some key differences:

- Application development projects depend on DevOps implementations. Explicit efforts must be made to minimize disruptions of Application development caused by DevOps changes.
- Processes and tools to support DevOps software changes may be different because of different programming languages and ecosystems.
- Integration of third-party tools is more prevalent with DevOps development.

- *Service Level Objectives (SLOs)* for the DevOps pipeline are typically different and much tighter than the SLOs for Application that depend on the DevOps pipeline.
- Coding practices for DevOps tooling are focused on the purpose of the tool rather than reliability of the Application. For example, a test script code is designed to find relevant failures in the Application, then stop, while the Application code is designed to continue to function as much as possible despite failures that may occur in a production environment.

This chapter explains how DevOps implementation projects are defined, including tasks, and how for each user story, product *Proof of Concept (POC)* trials are conducted to validate solution choices, the solution is validated with selected applications and use cases and deployed, training is conducted as the solution is deployed to production, and governance practices for the new solution are activated.

Why Is the Realize Step Important to Engineering DevOps?

This step is the actualization of the *DevOps Transformation Solution*. People are often impatient and tempted to skip the prior steps (Leadership, Alignment, Assessment, and Solution) and jump to this one. The failure to get alignment around a solution prior to implementation typically results in project failure because the leadership and team is not aligned behind the solution that is realized.

There are some key differences between DevOps projects and Application development projects that need to be considered in the Realize step of the Engineering DevOps Transformation. Application development depends on a highly stable DevOps environment. Changes to the DevOps tools and automation workflows can be disruptive to Application development if not implemented with the express purpose to minimize disruptions.

How Is the Realize Step Accomplished?

The Realize step has six substeps listed as follows and illustrated in **Figure 70—Realize DevOps Transformation Engineering Blueprint**:

- Task-Level Planning
- Proof of Concept Trials
- Implementation
- Release to Production
- Training
- Governance

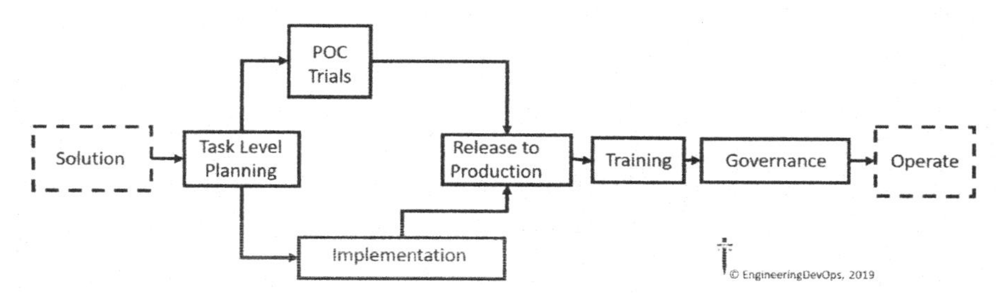

Figure 70—Realize DevOps Transformation Engineering Blueprint

Task Level Planning

Like Application development projects, tasks for the User Stories defined in the *DevOps Transformation Solution* are loaded into an *Agile* planning module *backlog*. For DevOps projects, it is critical to have a strategy for introducing change into the DevOps environment that will minimize disruption to Application developers. The DevOps Transformation *backlog* needs to prioritize tasks that are designed to minimize disruption. The following are some examples of tasks for minimizing disruption to Application development:

- Implement an isolated application environment for developing and testing DevOps changes.

- Identify Applications that can be used to validate the DevOps changes without undue risk.
- Deploy DevOps changes using a rolling release strategy in which the changes are gradually rolled out to the DevOps environments of Application development teams and rolled back if problems are found during usage.

Proof of Concept (POC) Trials

The tools chosen for the *DevOps Transformation Solution* are put to the test using *Proof-of-Concept (POC)* trails to validate they meet expectations and requirements against a select set of use cases. If two or more tools were chosen for comparison, then a "POC Bake-Off" may be required.

The following POC Bake-off process is recommended for comparing two tools. The same approach can be extended if more than two tools need to be compared. A methodical approach is recommended. Your team and the vendors will appreciate this. A good POC will return value to all stakeholders, as well as the vendors themselves—no matter which tool is selected.

Both tools can be made to fit the needs of your company. The POC is to analyze product features and evaluate the fit of each tool compared to what stakeholders expect the products need to do for them.

Define a list of requirements that cover the priority interests of all stakeholders. Ask all stakeholders to define their top two to three highest-priority tool requirements from their own point of view. Do not suppose you already know! By asking them, you will get their buy-in to the POC process, and most importantly, they are more likely to accept the results. Ask them to quantify the expected benefit as much as possible. Their answers will determine POC test cases and evaluation criterion.

- Business /executives: What features are the executives prepare to invest in? For example, reduce time to deploy, reduce wasted time setting up configurations, improve compliance audits, reduce maintenance costs, reduce security risks.

- Managers: What features will improve the job of the manager? For example, simplify work assignments, make work more predictable, easy to document, easier to train, establish recommended engineering practices, etc.
- Release Managers: What features will help release managers the most? For example, easy to document the configuration options have been tests and are supported by a release.
- QA staff: What features will help QA staff the most? For example, easy to validate all the configurations that are to be supported. Easier to orchestrate Green-Blue and A/B test strategies.
- Developers: What features will help developers the most? For example, simplify setting up a dev sandbox test environment, simplify pre-flight regression testing, being able to track configurations post-integration.
- Operations staff: What features will help operations staff the most? For example, reduce the effort to create and maintain configuration data, ease of automation and orchestration.
- Infra/tools staff: What features will help Infrastructure and tools staff the most? For example, reduce the drudgery and time to respond to requests to stand-up new configurations.
- Customers: What features will help customers the most? For example, reduced number of field failure events.

Prepare for the POC. Validate that there is a valid non-disclosure agreement in place with each vendor. Define the POC plan and review it with each vendor, including the following points:

- Business goals
- Test plan
- Test lab requirements (infrastructure, tools, licenses)
- Test data requirements
- POC schedule of events
- Vendor support requirements
- Documenting results plan

- Communication plan
- Explain how confidentiality will be assured between competing vendors
- Evaluation decision criterion

Ask each vendor if you missed anything in the plan that the vendor has seen before.

Define the test strategy, test cases, success criteria, and importance weights that will be used to verify each tool against it goals and to compare alternatives. Example test cases include the following:

- Documentation ease of use
- Ease of installation
- Support services
- Ability to replicate configurations
- Ability to audit configurations
- Ability to monitor configuration changes
- Ease of creating configuration recipes
- Roles and role-based controls
- Performance and capacity for deployment of large configurations
- High availability/fault tolerance
- Ability to test configurations before deployment
- Support for *Green-Blue* and *A/B* deployment strategies
- Ability to integrate with *Application Release Automation* tools such as ElectricFlow, Jenkins, etc.

The standard steps for this procedure are broadly described as follows:

1. Define a dedicated and equivalent test lab environment (systems and resources) that will be required for each tool. The lab doesn't have to be a full-blown lab; it can be a scaled-down environment, but there need to be tools or ways to test performance and capacity. One way is to use simulation tools or leverage temporary Cloud resources for those portions of the tests that require large numbers

of nodes. If the POCs for the tools are conducted in parallel, make sure confidential information between competing vendors is protected.

2. Set up the POC lab in such a way that it will be available and usable for the real project if the tool is chosen. For example, it could be left in place as a training environment after the POC. This way you will have a head start on the actual implementation. Review the test cases, weights and test lab resources with each vendor. Procure and prepare the lab infrastructure.
3. Bring the product in-house. Ask the vendor to assign an engineer to help with the POC. On-site work is strongly preferred because it increases the focus of the effort and reduces delays. Assign a dedicated resource to install and test the POC software for each tool. Ask the vendor to review and validate the installation before conducting tests.
4. Document everything. The documentation should not be limited to the lab and how it was configured. It should also include the steps of installation, results of tests that were performed, and final findings and recommendations.

   It's important to document the POC environment so it can be easily replicated for the teams supporting your product. Well-written documentation will also ensure that other teams within the organization understand the POC steps and why certain decisions were made.
5. Validate results with each vendor. As tests are run for each tool, report the results to the tool vendor. If there are any discrepancies, try to resolve them with the vendor as you go. Don't wait till the end. Document clearly any resolutions or revisions to the plan or workarounds that were used to resolve a discrepancy.
6. Compare results. Once all tests have been completed, prepare a comparison chart and call a meeting with internal stakeholders. This will provide data to drive towards a consensus recommendation.
7. Decide which tool to buy. This is critical. Make sure the business leaders are onboard with the decision.

8. Determine follow-up actions. Prior to discussion with vendors, list out what you will expect from the winning vendor going forward. Be sure to document a positive letter to both vendors and a high-level statement of the rationale for the decision. It is important to provide detailed feedback to the losing vendor regarding specific reasons they were not chosen. This must be done without disclosing confidential information about the competing vendor.
9. Communicate the decision. A meeting or detailed letter should be shared with internal staff immediately following the meetings with the vendors. A meeting with each vendor is recommended. Thank the vendor and present the letter. Go through the expected follow-up actions with the winning vendor.
10. Set follow-up actions dates. Agree with the winning vendor what the next steps and schedules are.

Implementation

DevOps changes should follow equivalent quality- and process-recommended engineering practices as code changes for Application development including design reviews, code reviews, test planning, code peer reviews, unit testing, functional and non-functional testing, integration, and regression testing. Where possible, it is recommended to apply DevOps practices to DevOps projects (i.e., "DevOps for DevOps").

The following are some of the recommended engineering practices for DevOps implementations:

- DevOps pipeline code, execution images and data needed for tools integrations, workflows, and environment orchestration are maintained in a multimaster repository with capabilities to roll back and roll forward to selected versions of pipelines.
- DevOps code changes should be supported with DevOps practices and have its own continuous delivery pipeline.
- New or changes to third-party tools need to follow vendor management and release acceptance practices similar to COTS

components of Applications.

- DevOps system changes is conducted in an isolated environment so as not to disturb the Applications that are using the production DevOps environment.
- Where possible, coding standards for DevOps tools leverage coding standards and tools for DevOps code and can leverage those used for Application code where there is overlap in programming languages.
- If the DevOps implementation uses different programming languages, then code quality checkers may be needed specific to the DevOps code.
- DevOps code often is integration code that is connecting different tools in the toolchain. Wherever possible, the DevOps code should be written as independent modules to avoid dependencies between modules. In the same context of microservices, the independent, modular code reduces the blast radius and reduces diagnostic and repair time when failures occur.
- Security practices for DevOps pipeline code should follow Application security practices such as code vulnerability scanning. Access to the large number of third-party tools used in DevOps toolchains is typically controlled using roll-based access controls and secrets vaults that dynamically assign temporary passwords. Sensitive customer data used in database testing needs to be obfuscated.
- Testing of DevOps implementation should follow *Continuous Testing* Practices including mostly automate functional and non-functional regression tests to ensure DevOps code changes do not break the trunk of DevOps code.
- Where possible, each version of DevOps pipeline code should be containerized.
- Containerized versions of DevOps pipeline code should be related to versions of Application and *Infrastructure-As-Code* so it is easy to roll-back or roll forward to known, tested combinations.
- It is possible to apply DevOps practices to DevOps projects (i.e., "DevOps for DevOps") and implement and deploy DevOps sys-

tem changes to the DevOps trunk incrementally using dark launch methods like Application releases.

Release to Production

Recommended engineering practices for releases of changes to DevOps code to production are very similar to Application code releases to production, except the target production environment is internal DevOps environments rather than live production. The best practice is to perform dark launches and rolling release strategies as described in this book.

It is critical to understand how DevOps implementation services are consumed by Application developers and align the release deployment process to the organization policies. Each DevOps pipeline variation that is supported should be kept in a Service Catalog so that access by Application developers is controlled.

Training

An organization may have multiple variations of pipelines to support different application stacks and deployment environments. Any DevOps pipeline may have tens of tools in the toolchain as needed for development, version management, builds, testing, integrations, monitoring, security, infrastructure orchestration, management, governance, automation, and delivery/deployments. It is a goal of well-engineered DevOps solutions to hide complexity from Application developers so they can efficiently develop application improvements instead of worrying about all the intricacies of the DevOps toolchain and infrastructures. Despite this goal, there are stakeholders that need to know different slices of information of the DevOps solution, and training is needed for each of them, detailed as follows:

- Business leaders need to know how to monitor progress on DevOps projects and how to monitor release performance of business applications.

- Application developers needs to know how to use the DevOps pipeline and tools for building, testing, debugging and deploying their code changes.
- DevOps pipeline administrators need to know how to monitor performance and diagnose and repair DevOps pipeline components.

As new components or changes to existing components of a DevOps solution are implemented, training needs to be developed at the same time and be available to support deployment and operations of the DevOps solution.

Governance

The primary concerns regarding governance of DevOps pipelines are as follows:

- Control costs. The use of elastic infrastructure resources such as cloud capabilities can be expensive if resource allocations and hold times are not controlled.
- Avoid sprawl of too many DevOps pipeline variations. While there are legitimate reasons for different variations (e.g., application stacks and deployment environments), too many variations increase maintenance costs and reduced the opportunity for sharing between applications.
- Provide operational visibility. Ensure all pipeline variations have common monitoring access points (agent based or agentless does not matter) for stakeholders to observe the use and performance of each DevOps pipeline against SLIs, SLOs, and SLAs.
- Implement security. Each pipeline variation that is supported needs to have a minimal set of security features including role-based access controls, limitations of using only trusted components, and data obfuscation.
- Guarantee release quality and reliability. Each continuous delivery pipeline needs release policies to validate release quality and

reliability metrics are within prescribed thresholds prior to release. The metrics may include SLOs agreed with SREs.
- Support controlled experimentation by limiting access to experimental DevOps pipeline solutions.

Popular approaches for implementing Governance controls include the following:

- Use cloud management tools to control resource cost utilizations.
- Use Service Catalogs to limit the number of DevOps pipelines available to users.
- Each DevOps pipeline has embedded monitoring tools to support operations visibility.
- Security tools are built into each DevOps pipeline variation.
- Application Release Automation tools implement stage-by-stage gate control policies as code.
- Value-Stream Management tools implement overall end-to-end governance as code policies.
- Service Catalogs support roll-based access controls.

Overcoming Challenges with the Realize Step

Some of the most prevalent challenges and mitigations with the Realize step are listed in this section.

Application development for organizations that do not follow a disciplined DevOps pipeline solution implementation process suffer from unreliable, error-prone, inconsistent Application release performance and generally do not achieve the goals. There is a risk that the entire environment will revert to an immature chaos state.

Developing DIY tools rather than engaging Open-Source or Enterprise tools can be a trap that is hard to recover from. Most often DIY tools cost more to build and maintain, are less scalable, and are less evolvable than the creators will admit. Unless there is a proprietary reason to develop DIY tools, they should be avoided.

DevOps pipeline development without strong governance can quickly mushroom into too many DevOps pipeline variations, cost overruns, and security problems. Every DevOps Transformation project needs to consider governance requirements.

25

Step Six: Operationalize

Once a DevOps solution is deployed, it doesn't take care of itself! It requires management. Deployed DevOps solutions are available under controlled circumstances, monitored, governed, supported, expanded, and evolved. All these things require processes and human involvement.

Why Is the Operationalize Step Important to Engineering DevOps?

Access to DevOps solutions needs to be convenient for Application developers, yet controlled so that only authorized users have access.

Until something becomes a habit, it will not be used. Monitoring is needed to ensure DevOps solutions that are deployed to Application developers are used and that they are effective.

SLAs for DevOps solutions are typically different that the SLAs for operational infrastructure for Applications.

User support is essential to ensure users have access to information and can report problems and enhancement requests.

As a DevOps solution is used by more and more Application developers, additional capacity will need to be provisioned.

No useful system remains static. Provisions for controlled evolution of the DevOps solution are required.

How Is the Operationalize Step Accomplished?

Operating a DevOps Solution that has been deployed for use by Applications developers has similar requirements to operating infrastructures for Applications. The key difference is the focus of what is being operated. For DevOps operations, it is the DevOps solution rather than the Application that is being developed that is the focus of operations. As illustrated in **Figure 71—DevOps Solution Operations Engineering Blueprint**, the following are the major components of DevOps operations:

- Controlled Access
- Monitoring
- Governance
- Support
- Evolution

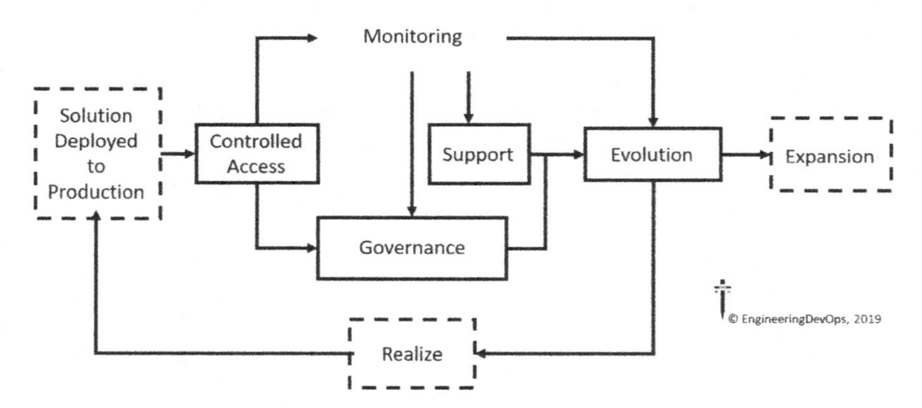

Figure 71—DevOps Solution Operations Engineering Blueprint

Controlled Access

Each variation of a DevOps pipeline may be statically configured or instantiated on-demand. Access to each variation of a DevOps pipeline solution will be controlled to ensure only authorized stakeholders can monitor or use it. This is typically accomplished with role-based access controls and Service Catalogs.

Monitoring

Usability of a DevOps solution is monitored to see how each type of stakeholder is using it and to keep track of problems reported through the Support system.

Site Reliability Engineering (SRE) practices monitor *SLI*, *SLO*, and *SLA* metrics of the DevOps solution. The *SLAs* for the DevOps solution are typically different that the Application. DevOps solutions for organizations that can tolerate failures to the DevOps environment may have more relaxed *SLOs*, while an organization that cannot tolerate failures to the DevOps environment may have tighter *SLOs* for DevOps than the Application.

Value-Stream Management solutions keep track of end-to-end activities to ensure they are not exceeding thresholds. For example, lead times for each stage in the pipeline may have a budget to ensure *Continuous Flow*. *VSM* can track actual lead times per DevOps stage, flag exceptions for review, and include results in a trend analysis.

Governance

Cost controls provided by cloud management solutions ensure users do not stand up or retain costly resources excessively, according to governance policies set by the organization.

Resource utilization trends are used for capacity planning. When the trends indicate resources are being exhausted too frequently acquisition, of additional resources may be scheduled.

Support

Users report questions or enhancement requests for deployed DevOps solutions through a ticket system. The tickets are used to track replies and to provide input to the evolution process. Examples of user requests may be changes to access privileges, directions to training information, tool user manuals, or governance policies. Examples of enhancement requests may be request for new DevOps pipeline configurations, new tool stacks, new tools, new workflows, or suggestions to enhance training information.

Evolution

DevOps solutions that are exclusively serving Applications that rarely change, such as legacy systems, do not need to be evolved unless a component become obsolete and needs to be replaced.

Almost all other DevOps solutions need to be evolved to keep up with changes in scale, capacity, compatibility with upgraded tool stacks, and new features required for new Application requirements.

The primary source for evolution priorities come from users in the form of tickets. In addition, *Retrospectives* are conducted after each release of a DevOps solution to create actionable prioritized lessons learned for continuous improvement.

Overcoming Challenges with the Operationalize Step

- If some of the tools in the DevOps toolchain do not support auto-scaling, they may need to be replaced with tools that do as the solution expands.
- If *SLAs* for the DevOps solution are not determined and agreed with users, there can be serious misunderstandings between operations and the users of the tools. Work out an agreed *SLA* as soon as possible and put in place monitoring metrics to detect exceptions and procedures to action exceptions.
- If users are requesting changes that violate *governance policies*, then a new DevOps solution may need to be worked out.

26

Step Seven: Expansion

Once *Continuous Flow* (*The First Way* of DevOps) is realized for a select set of applications, the organization can safely proactively share recommended engineering practices (*Yokoten*), expand proven solutions to other applications, pipeline variations, and deployment regions across the organization. Further evolution cycles will lead to mastery of *Continuous Flow*, *Continuous Feedback* (*The Second Way* of DevOps) and *Continuous Improvement* (*The Third Way* of DevOps). Once *Continuous Improvement* is mastered, the organization will be well positioned to take advantage of next generation DevOps solutions—a future level I call ***Continuous Autonomous Improvement.***

Why Is the Expansion Step Important to Engineering DevOps?

The realization of DevOps across an enterprise with multiple types of Applications, pipeline variations, and deployment infrastructures is akin to hunting a moving target, because all of the *Three Dimensions of DevOps* (People, Process, and Technology) are constantly changing. The engineering approach to converge the enterprise to a common goal, despite the differences and shifting, is to establish a strong reliable reference, or model implementation, from which other applications can learn as it is evolved through each level of maturity.

In large organizations, there is tendency to let multiple Applications evolve DevOps solutions in parallel and be alternative models for other Applications. This can also work and may achieve early success faster for a larger number of Applications, but most likely it will result in disparate competing solutions that become harder and harder to converge towards an enterprise solution at higher levels of DevOps maturity.

How Is the Expansion Step Accomplished?

Figure 72—DevOps Expansion Engineering Blueprint illustrates an engineering DevOps expansion blueprint that includes the following components:

- DevOps *Continuous Flow* Model Evolutions
- DevOps *Continuous Flow* Mastery
- *Yokoten*—Proactive Sharing
- Project Approvals for Expansion Projects
- DevOps Maturity Level 4—*Second Way Continuous Feedback* Evolutions
- DevOps Maturity Level 5—*Third Way Continuous Improvement* Evolutions
- Beyond DevOps Maturity Level 5—*Continuous Autonomous Improvement*

The blueprint shows a progression of DevOps to higher levels of maturity from left to right with a continuous loop of realized improvements, with operations experience of each improvement driving the next evolution. At each level, the *Continuous Flow* model Application is the focus for improvements, and other applications proactively share lessons learned from experiences with the model through a process called Yokoten. At the higher levels of maturity, improvements are tested for performance impacts to *Continuous Flow*. The goal is to achieve mastery of Maturity Level 5, *Continuous Improvement*. Beyond *Continuous Improvement*, it is speculated in the future that the continuous improvements will be advanced by technologies to such an extent they will be autonomous.

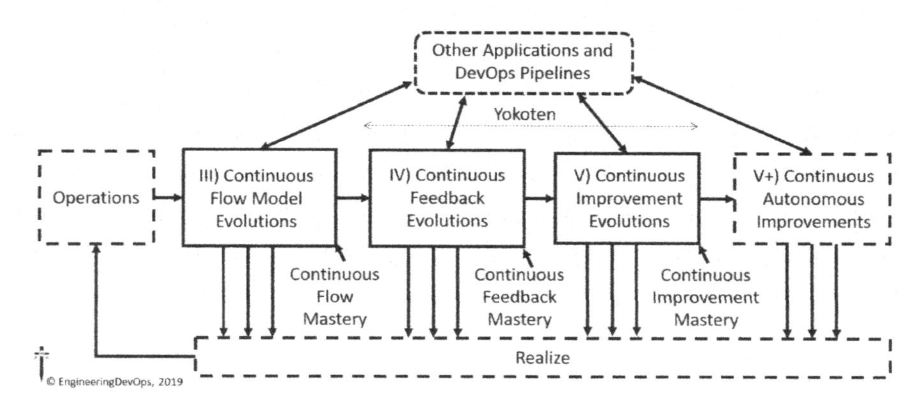

Figure 72—DevOps Expansion Engineering Blueprint

DevOps Continuous Flow Model Evolutions

Continuous Flow is achieved when all the stages of the *CI/CD pipeline* are defined and the *gate criteria* for promoting or rejecting changes between each stage are in place. For less-mature *Continuous Flow* pipelines, some of processes for each stage and gate decisions may be manual and have significant bottlenecks. Each evolution of the *Continuous Flow* implementation works towards removing bottlenecks in the flow, primarily through process and decision automation.

DevOps Continuous Flow Mastery

DevOps *Continuous Flow* Mastery is accomplished when *Continuous Flow* is fully autonomous. A good change committed by a developer will propagate all the way through the CI/CD pipeline with no manual intervention. A change that is not good will automatically be rejected and reported, and the most recent good combination of the application code, pipeline, and infrastructure will be restored automatically.

Yokoten—Proactive Sharing

As indicated by Al Norval with the firm Lean Pathways, Inc., "*Yokoten* is a process for sharing learning laterally across an organization. It entails copying and improving on ideas that work."

You can think of *Yokoten* as *"horizontal deployment"* or *"sideways expansion."* The corresponding image is one of ideas unfolding across an organization. *Yokoten* is horizontal and peer-to-peer, with the expectation that people go see for themselves and learn how another area did Kaizen and then improve on those *Kaizen* ideas in the application to their local problems.

It's not a vertical, top-down requirement to "copy exactly." Nor is it a "recommended engineering practices" or "benchmarking" approach, nor is it what some organizations refer to as a "lift and shift" model. Rather, it is a process where people are encouraged to go see for themselves and return to their own area to add their own wisdom and ideas to the knowledge they gained.[RW44]

The *Yokoten* concept is extremely important to DevOps expansions. The model application and pipeline serve as a key reference implementation that other applications and pipeline variations can proactively learn from. Conversely, experiences of other application can be shared with the model application and pipeline to ensure the reference implementation represents the organization's best practice.

Mastering DevOps Maturity Level 4—Second Way (Continuous Feedback) Evolutions

The Second Way of DevOps, *Continuous Feedback*, involves evolutions that enhance visibility of the Application, pipeline, and infrastructure components in ways that are more and more effective until *Continuous Feedback* is mastered.

The following are some practices indicative of *Continuous Feedback* mastery:

- SLIs, SLOs, and SLAs are in place to monitor and measure the health and performance of the Application, the pipeline, and the infrastructure.
- Continuous Flow automatically makes use of the SLIs to control flow throughout the end-to-end CI/CD pipeline.
- Release decisions are completely automated with policies that make use of SLIs, SLOs, and SLAs.
- *Continuous Security* metrics are automatically rolled up to a security management system that validates security thresholds are being met and, if they are not, flags them for resolution.
- The end-to-end value stream from planning through to operations is instrumented with metrics that are automatically rolled up to a value-stream management system that analyzes trends.
- Governance processes are driven by data automatically.
- Training programs emphasize using data for decision-making.

Mastering DevOps Maturity Level 5—Third Way (Continuous Improvement) Evolutions

The Third Way of DevOps, *Continuous Improvement*, involves evolutions that enhance the ability to experiment and evolve the Application, pipeline, and infrastructure components in ways that are more and more effective until *Continuous Improvement* is mastered.

The following are some practices indicative of *Continuous Improvement* mastery:

- The organization proactively is scanning the industry for relevant DevOps innovations.
- Incentives and reward systems drive staff to experiment with innovative changes to the Applications, pipeline, and infrastructure.
- Recovery processes are automatically triggered when changes to the pipeline or infrastructure cause problems.
- Retrospectives are triggered by analysis derived from *Continuous Feedback* mechanisms.

- Governance processes offer controlled flexibility so teams can experiment with novel variations of Applications, pipelines, and infrastructures in well-controlled environments that contain risks.
- Internal DevOps training programs emphasize experimentation, continuous improvements, and risk mitigation methods.

Beyond DevOps Maturity Level 5—Continuous Autonomous Improvement

Nils Bohr, Nobel laureate in physics, once said, “Prediction is very difficult, especially if it’s about the future.” Arthur C. Clarke said, “Any sufficiently advanced technology is indistinguishable from magic.” With these two distinguished opinions in mind, we will explore the future of DevOps beyond Maturity Level 5 in the next chapter. My own view is that current state-of-the-art technologies such as Artificial Intelligence and Machine Learning will soon become state-of-the practice for at least several DevOps applications. Once that occurs, I imagine a technology breakthrough for DevOps, one in which humans can be removed from the *Continuous Improvement* loops of DevOps Maturity Level 5 that today require them to interpret analysis and prescribe solutions—at least for a number of applications that today are chronic sources of bottlenecks such as infrastructure orchestration, testing, and release deployments.

Overcoming Challenges with the Expansion Step

The primary challenge organizations face with the Expansion step is to keep DevOps evolution a priority as it matures. DevOps is not natural. As this book has explained, accomplishing any level of DevOps maturity requires a strategy; alignment across teams; and hard, disciplined work. DevOps does not exist as an island. Other projects are competing for the same investment dollars. They key is to remind business leaders that while DevOps is not an island; it is a river that runs through all things in business innovation and operations, and it is a critical factor affecting competitive capabilities for the business. Fail at DevOps and risk business failure.

27

Future of Engineering DevOps—Beyond *Continuous Improvement*

"Then Sir Bedivere departed, and went to the sword, and lightly took it up, and went to the water side; and there he bound the girdle about the hilts, and then he threw the sword as far into the water as he might; and there came an arm and an hand above the water and met it, and caught it, and so shook it thrice and brandished, and then vanished away the hand with the sword in the water."

"Hic iacet Arthurus, rex quondam, rexque futurus" or "Here lies Arthur, king once, and king to be."

"Yet some men say in many parts of England that King Arthur is not dead, but had by the will of our Lord Jesu into another place; and men say that he shall come again, and he shall win the holy cross."

—**Le Morte d'Arthur**, *BOOK XXI CHAPTER V-VII*

DevOps implementations and DevOps itself are not immortal. As DevOps leaders, experts, and evangelists come and go, the perceived value of DevOps may wax and wane. The fame of innovative new technologies expires ever more quickly with the acceleration of technological systems facilitated by stacking technologies higher and higher. Today's state-of-the art is tomorrow's state-of-the practice. But what is certain is tomorrow will come, and DevOps will continue to evolve. What is important now is for you to do what you can to be ready.

"DevOps has quickly matured from a fairy tale to a bestseller at most large enterprises. However, we are still at the beginning of the DevOps story, with many exciting chapters still to be written."

—**Mark Campbell**, Chief Innovation Officer, Trace3

How will new, potentially disruptive technologies impact the future of DevOps? When pondering the future of DevOps, it is useful to bear in mind *The Three Dimensions of DevOps* (People, Process, and Technology), *The Nine Pillars of DevOps* (Leadership, Collaborative Culture, Design for DevOps, Continuous Integration, Continuous Testing, Elastic Infrastructure, Continuous Monitoring, Continuous Security, and Continuous Delivery), and the *Twenty-Seven Critical Success Factors for DevOps* that are a cross-product of the dimensions and pillars. **Figure 73—DevOps Beyond *Continuous Improvement* Engineering Blueprint** presents ten key technologies that are influencing the evolution of DevOps itself.

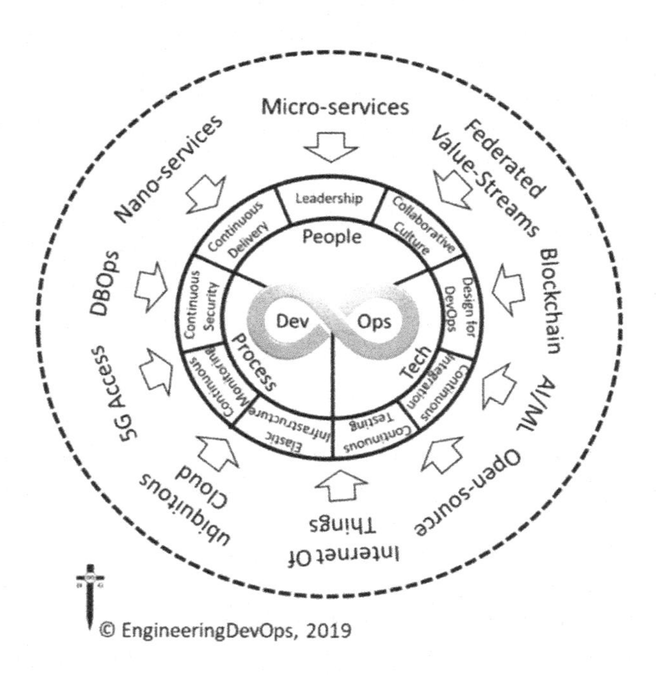

Figure 73—DevOps Beyond Continuous Improvement Engineering Blueprint

The ten key technologies shown in the Engineering DevOps Beyond *Continuous Improvement* Blueprint are as follows:

1. **Microservices**—The value Applications build onto microservices architectures has been proven. Microservices design practices have already moved from state-of-the- art to state-of-the practice. Most enterprises that have legacy monolithic systems are either replacing them or refactoring them.[RB11, RW51]
2. **Nanoservices**—Software services that can be used by microservices such as Functions-as-a-Service and Serverless offerings by cloud service providers are currently becoming more mainstream.[RW50]
3. **Federated Value-Streams**—This refers the need to manage multiple pipelines to deliver and support changes to applications composed of a portfolio of microservices pipelines. This requirement is tied to the evolution of microservices and increasingly served by Value-Steam Management solutions that preside over a portfolio of CI/CD pipelines.[RW52]
4. **Blockchain**—This provides a trustful network for communications between distributed processes. These processes can be for applications, pipelines, or infrastructures.[RW45]
5. **AI/ML**—Artificial Intelligence and Machine Learning apply algorithms and heuristics to data analysis problems. DevOps generates a lot of data to characterizes performance of Applications, pipelines, and infrastructures. The analysis of the data can be overwhelming to humans, even when it is presented in processed trend categories. AI/ML applications in DevOps are already appearing in channelling, communication operations data analysis, and test results analysis to identify anomalous behaviors and bottlenecks that can be targeted for improvement.[RW46]
6. **Open-Source Software**—The trend is that more and more application, pipeline, and infrastructure software are open-source. The percent of code that is enterprise is reducing. I predict that nearly all software will be open-source within a five- to ten-year planning horizon.[RW47]

7. **Internet-of-Things (IOT)**—If you think about IOT as the distribution of a huge number of smart (software-driven) hardware devices to the very edge of network (mobile and on-premises), you can understand that continuous delivery and deployment and security for those devices need to be capable of handling massive parallelism.[RW48]
8. **Ubiquitous Cloud Computing**—The movement to cloud computing is well underway in most enterprises. Enterprises want their infrastructures to seamlessly use multi-cloud so they can move workloads to any cloud depending on commercial factors such as price. The desires of developers of applications and pipelines are expected to drive cloud service providers to offer more and more equivalent capabilities as other cloud service providers, even as they continue to offer innovative differentiating capabilities.[RW49]
9. **5G Access**—Very high-speed wireless access is pushing up the speed of computing collaboration across wide area networks. The reduced latency offered by 5G is enabling near-real-time infrastructure configurations. 5G opens new DevOps possibilities for more distributed application, pipelines, and infrastructure; reduces bottlenecks; and also offers challenges for the current generation of DevOps systems to keep up with the blazing fast speeds.
10. **DBOps**—Solutions that address analysis of extreme data volumes across an ever more distributed computing network is driving the need for the integration of DevOps data pipelines correlated in step with application code pipelines.[RW45]

So where are these new technologies taking DevOps? When and how can you prepare yourself and your DevOps for the future?

My crystal ball is telling me a few things about the likely impacts of these new technologies on application, pipelines, and infrastructures.[RW68]

Applications are evolving to leverage the capabilities of microservices, nanoservices, block-chains, 5G speeds/low latency, IOT devices, open-source, and distribute databases. This combination is conspiring to make applications more and more distributed and dependant on highly secure, high-bandwidth, low-latency networks that can host their workloads and serve as their DevOps delivery platforms.

Pipelines must evolve to keep up with the high-capacity, highly parallel, and distributed application deployment architectures. Federated value-stream management systems are become more important. DevOps tools must be deployable close to the distributed application and therefore must run equally well on any cloud platform. The volume of data is growing. Pipeline bottlenecks that are exacerbated by a distributed architecture, such as communications, analysis, and testing, must leverage AI/ML to keep up. Open-source DevOps tool solutions are on the rise. The rapid rise of open-source containers and Kubernetes are a harbinger of the future of DevOps tools.

Infrastructures are becoming even more elastic. The rapid rise of ubiquitous cloud computing, 5G access, and IOT indicate infrastructures are moving from software configurable systems at a macro device level to a more distributed microlevel. Alone with this highly distributed ephemeral infrastructure paradigm, new security attack surfaces will need to leverage highly secure and distributed blockchain technology.

So what will the next level of DevOps maturity beyond *Continuous Improvement* look like? At all times, DevOps aims to remove bottlenecks. The new technologies introduce new challenges that could add new bottlenecks, and they also offer new opportunities to reduce bottlenecks. DevOps solutions must evolve to support pipelines and orchestrated infrastructures for ever more distributed Applications. The current generation of DevOps tools that were mostly designed for integration into serialized toolchains and separate unfederated pipelines per application could become a bottleneck. The next generation of DevOps tools must be not only cloud-ready. They must be cloud-optimized and configurable into highly federated, networked toolchain topologies.

AI/ML are already showing applications to reduce bottlenecks that come from highly networked communication, data analysis, and testing structures. It is expected that data generated by DevOps solutions will be consumed by AI/ML algorithms that will provide insights on how to reduce existing bottlenecks and prevent new networked structures from adding new bottlenecks.

It is debatable whether such improvements would constitute the beginnings of a new level of DevOps maturity or really be considered within the scope of existing *Continuous Improvement* Maturity Level 5.

If we look further into the future, the prospect of eliminating the biggest bottleneck—human interactions—from DevOps pipelines could result in a new DevOps Maturity level 6, which I would refer to as *"Continuous Autonomous Improvement."* DevOps' use of AI/ML is replacing the need for humans to some extent for a limited number of mostly passive analysis applications already. As AI/ML is applied in a more active capacity, such as automating test creation, automating build schedules, and automating infrastructure orchestrations, then it would be feasible to talk about mastery of a DevOps Maturity Level 6.

28

Continuous Learning

DevOps is an evolving body of knowledge. There is no one-stop shop that covers all roles and the complete range of skills required for DevOps. A DevOps-competent workforce requires an ongoing commitment to DevOps skills development and mastery. Enterprises must approach DevOps training strategically to leverage training resources to best advantage. **Figure 74—DevOps Continuous Learning Engineering Blueprint** shows the primary sources for DevOps knowledge.

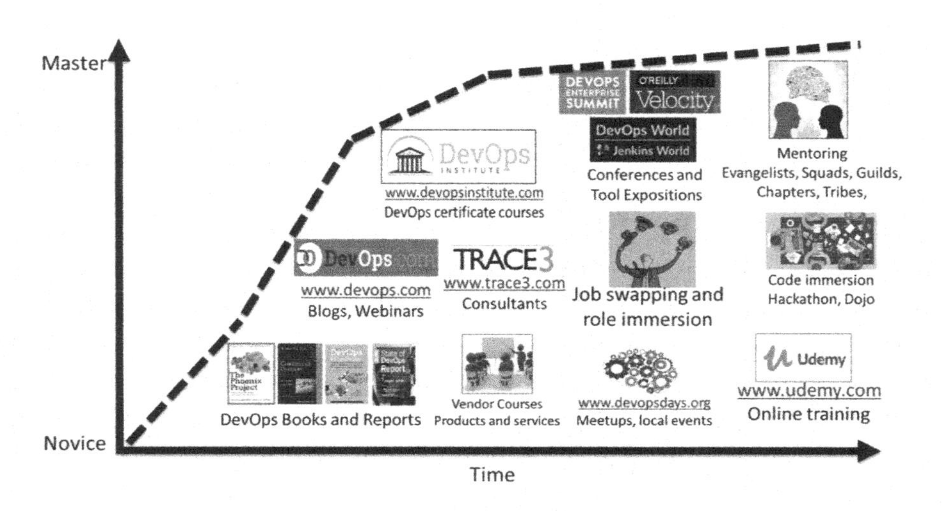

Figure 74—DevOps Continuous Learning Engineering Blueprint

Practices and tools to implement the DevOps Continuous Learning Blueprint are explained in the following sections.

Learning Continuous Flow

The First Way of DevOps, *Continuous Flow*, is the foundation for DevOps.

Why Is Learning *Continuous Flow* Important to Engineering DevOps?

- A knowledgeable, skilled workforce is needed to meet strategic objectives.
- Without a solid understanding of DevOps concepts and terminology, communication between people and teams is inefficient and a source of bottlenecks and misunderstandings.
- Developing hands-on skills for the tools and workflows is needed to understand and implement DevOps pipelines.
- There is an industry-wide shortage of skilled DevOps candidates.
- New skills foster innovation.
- Knowledge workers reduce production costs and reduce mistakes.
- Having skilled coworkers makes for a satisfying working environment.
- Employee retention is better.
- Competitive advantages are derived from knowledge.

How Is Learning *Continuous Flow* Accomplished?

The following are recommended engineering practices for learning *Continuous Flow:*

- People are encouraged to be continuous learners with an ongoing commitment to DevOps skills development.
- DevOps training is approached strategically to ensure DevOps training is well aligned to business goals.

- DevOps skills gap assessments are conducted and used to define and refine the training strategy.
- Leaders and cross-functional teams are encouraged to take DevOps Fundamentals training.
- Leaders define a training strategy that best fits the needs and priorities of the organization to meet critical skills gaps.
- The training strategy considers which skills and roles can be covered by inexpensive self-study resources. Which skills and roles warrant more costly training?
- DevOps training has a distinct budget.
- Compliance to DevOps training programs are tracked.
- DevOps leadership training is offered to management and staff that have a role relevant for DevOps.
- DevOps tools training includes the recommended engineering practices for using each DevOps tool in the Service Catalog.
- Staff are encouraged and supported to take DevOps certifications that are relevant to their assigned roles.
- A DevOps training program is defined and includes a recommended curriculum for each role including leaders, developers, QA, Ops, and infrastructure and security staff that are operating in a DevOps environment.
- Recommended DevOps training resources are cataloged, and the catalog is visible for all stakeholders.
- Merit and recognition programs encourage employees to master DevOps skills.
- Time for taking DevOps training is allocated for all staff.
- DevOps Mentoring programs supplement formal training.
- Job swapping and immersive training is encouraged and planned.

People are encouraged to attend external conferences and meetups and share what they learn with other team members.

What Is Needed to Implement Learning for *Continuous Flow?*

The following are recommended engineering practices:

- Assign roles and responsibilities for DevOps training leadership (e.g., DevOps Sensei, Training PM, evangelist).
- Conduct a DevOps skills gap assessment.
- Define a training strategy that best address the gaps and fits the needs and priorities of the organization.
- Define a training budget that includes expenses and time allocations per role and skill. Which skills and roles can be covered by inexpensive self-study resources? Which skills and roles warrant more costly training?
- Identify training goals for each role.
- Ensure recruiting job profiles match skill requirements for DevOps skills that are most relevant to the organization strategic goals.
- Put in place means to make training resources visible and available.
- Set training goals for teams and individuals.
- Put in place incentives and recognition programs for training.

Learning *Continuous Feedback*

There are few training resources dedicated solely to *The Second Way* of DevOps—*Continuous Feedback*. Many of the same resources as *The First Way* of DevOps have some relevant information. DevOps monitoring, release automation, and security tools publications, articles, and tool vendors are a good source. A DevOps expert experienced with mature DevOps can select the training information that is relevant to the specific *Continuous Feedback* tools and feedback metrics used by the enterprise.

Why Is Learning *Continuous Feedback* Important to Engineering DevOps?

The Second Way of DevOps—*Continuous Feedback*—requires a mature understanding of DevOps concepts and practices.

Specific skills are needed for implementation and operation of advanced monitoring tools and practices for Application performance, Database performance, advanced continuous testing, advanced release management, infrastructure performance, and continuous security.

How Is Learning *Continuous Feedback* Accomplished?

The following are recommended engineering practices for learning *Continuous* Feedback:

- Roles and responsibilities are assigned to experts for defining a training program for DevOps *Continuous Feedback* (e.g., DevOps Sensei, Training PM, evangelist).
- *Continuous Feedback* DevOps skills gap assessments are conducted periodically.
- Dev and Ops training strategies are designed to address gaps and priorities of the organization for DevOps *Continuous Feedback.*
- A training budget that includes allocations of expenses and time allocations per role exists to support *Continuous Feedback.*
- *Continuous Feedback* goals and SLAs are defined for each role.
- Recruiting job profiles match skill requirements for DevOps skills relevant for *Continuous Feedback.*
- Training resources are available and visible to those that need *Continuous Feedback* training.
- Training goals set for teams and individuals include goals for *Continuous Feedback.*
- Incentives and recognition programs support *Continuous Feedback* training.
- *Continuous Feedback* training progress is tracked and visible.

What Is Needed to Implement Learning for *Continuous Feedback?*

- Assign roles and responsibilities to experts for defining a training program for DevOps *Continuous Feedback* (e.g., DevOps Sensei, Training PM, evangelist)
- Conduct a *Continuous Feedback* DevOps skills gap assessment.
- Define a training strategy that best address the gaps and fits the needs and priorities of the organization.
- Define a training budget that includes expenses and time allocations per role and skill. Which skills and roles can be covered by inexpensive self-study resources? Which skills and roles warrant more costly training?
- Identify training goals for each role.
- Ensure recruiting job profiles match skill requirements for DevOps skills that are most relevant to the organization's strategic goals.
- Put in place means to make training resources visible and available.
- Set training goals for teams and individuals.
- Put in place incentives and recognition programs for training.
- Track training progress.

Learning *Continuous Improvement*

There are few training resources dedicated solely to *The Third Way* of DevOps—*Continuous Improvement*. However, many of the same resources as *The First Way* and *Second Way* of DevOps have some relevant information. Microservices and SRE publications and articles and related tool vendors are a good source. An expert with experience with mature DevOps and SRE knowledge can select the training information that is relevant to the specific continuous feedback tools and continuous feedback metrics used by the enterprise.

Why Is Learning *Continuous Improvement* Important to Engineering DevOps?

The Third Way of DevOps—*"Continuous Improvement"*—requires an organizational commitment to continuous learning.

Specific skills are needed for implementation and operation of advanced microservices, multipipelines, and SRE.

Retention and development of highly skilled and hard to find experienced DevOps and SREs makes it important to develop skills in-house.

How Is Learning *Continuous Improvement* Accomplished?

The following are recommended engineering practices for learning *Continuous Improvement:*

- A continuous learning culture encourages individuals to continuously gain knowledge and exchange experiences and ideas for DevOps and SRE.
- Mentoring is an explicit program.
- Retrospectives are structured to thoroughly analyze significant unexpected events and determine solutions that will prevent the events in the future.
- An active online library of DevOps and SRE training resources is maintained that tracks the latest and most innovative approaches to DevOps and SRE.
- DevOps and SRE teams evolve their skills and advance their careers by engaging with industry thought leaders to continuously update training content.
- Partner with trusted education and training partners ensure the highest-quality learning content and tools are made available to DevOps and SRE teams.
- A portion of DevOps and SRE teams time is allocated for training and education.
- Cross-team sharing of recommended engineering practices is proactively pursued using the concepts of Yokoten—proactive sharing of practices between teams.

What Is Needed to Implement Learning *Continuous Improvement?*

- Create a continuous learning culture that encourages individuals to continuously gain knowledge and exchange experiences and ideas for DevOps and SRE.
- Build an active online library of DevOps and SRE training resources that tracks the latest and most innovative approaches to DevOps and SRE.
- Engage with industry thought leaders to continuously update content for DevOps and SRE teams to evolve their skills and advance their careers.
- Partner with trusted education and training partners to ensure the highest-quality learning content and tools are made available to DevOps and SRE teams.

 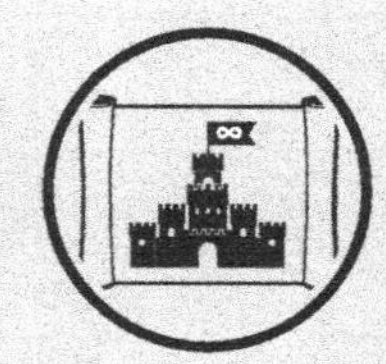  

PART V

APPENDIX AND REFERENCES

Merlyn took Arthur's hand and said kindly, "You are young, and do not understand these things. But you will learn that owls are the most courteous, single-hearted and faithful creatures living. You must never be familiar, rude or vulgar with them, or make them look ridiculous. Their mother is Athene, the goddess of wisdom, and, although they are often ready to play the buffoon to amuse you, such conduct is the prerogative of the truly wise."

—**T.H. White**, *The Once and Future King*

Athene is the ancient Greek goddess of wisdom, courage, inspiration, civilization, law and justice, strategic warfare, mathematics, strength, strategy, the arts, crafts, and skill. Mastery of DevOps is a continuous quest for knowledge and skills across all dimensions and pillars. The quest for learning is without end. You will benefit if you connect with mentors with deep experience. In this way, consider yourself a humble child of knowledge. Be forever inspired by the wise, and faithfully engage in the joy of learning, even when the material appears at first obscure or absurd. Be generous with what you learn. Mentor and inspire others.

PART V: Engineering DevOps Appendices, and References includes materials and sources that are used for engineering DevOps.

These appendices contain information that is referred to in *Engineering DevOps*. It includes templates and checklists that are used for the DevOps Seven-Step Transformation Engineering Blueprint. Many of these items also appear in Microsoft Word, PowerPoint, or Excel format on my website:

www.EngineeringDevOps.com

Appendix A

Definition of DevOps Engineering Terms

Appendix A of this document provides an alphabetized list of the most popular DevOps terms, abbreviations, and acronyms, with a description consistent with their use in this document. For a more comprehensive list of DevOps definitions, refer to the "Definitions of DevOps Terms" document posted on my website (www.EngineeringDevOps.com).

Application Release Automation (ARA)—The processes of packaging and deploying an application or update across the complete release life cycle, ultimately enabling new products and services to be brought to market faster.

Application Under Test (AUT)—The AUT is a software application.

Artifact—Any element in a software development project including documentation, test plans, images, data files, and executable modules.

Artifact Repository—Store for binaries, reports, and metadata for each release candidate (e.g., Archiva, Nexus, JFrog).

Behavior-Based Testing—Test cases are created by simulating an AUT's externally observable inputs.

Blue/Green testing—Taking software from the final stage of testing to live production using two environments labeled Blue and Green. Once the software is working in the Green environment, switch the router so that all incoming requests go to the Green environment—the Blue one is now idle.

CAB—Change Advisory Board.

CALMS—Pillars or values of DevOps: Culture, Automation, Lean, Measurement, and Sharing.

Canary Testing—A canary (also called a canary test) is a push of code changes to a small number of end-users who have not volunteered to test anything.

Chat-Ops—An approach to managing technical and business operations through a group chat room (e.g., Slack).

Check-In—Action of submitting a software change into a system version management system.

CI/CD—see Continuous Integration, Continuous Delivery, and Continuous Deployment.

Cloud-Native—Native cloud applications (NCA) are designed for cloud computing.

Clustering—A group of computers (called nodes or members) work together as a cluster connected through a fast network acting as a single system.

CMDB—Configuration Management Database.

Configuration Management (CM)—A system engineering process for establishing and maintaining consistency of a product's performance, functional, and physical attributes with its requirements, design, and operational information throughout its life.

Containers—Containers wrap up a piece of software in a complete file system that contains everything it needs to run: code, runtime, system tools, system libraries—anything you can install on a server. This guarantees that it will always run the same, regardless of the environment it is running in (e.g., Docker, Valgrant).

Continuous Delivery (CD)—A software development discipline where you build software in such a way that the software can be released to production at any time.

Continuous Delivery Architect—A person who is responsible to guide the implementation and recommended engineering practices for a continuous delivery pipeline.

Continuous Delivery Pipeline—A series of processes that are performed on product changes in stages. A change is injected at the beginning of the pipeline. A change may be new versions of code, data, or images for applications. Each stage processes the artifacts resulting from the prior stage. The last stage results in deployment to production.

Continuous Delivery Pipeline Stage—Process step in a continuous delivery pipeline. These are not standard. Examples are design (determine implementation changes), creation (implement an unintegrated version of design changes), integration (merge created changes into a version of a product), building (produce a version of a product subsystem), binding (produce a version of the product system into package artifacts), delivery (produce a release candidate version of artifacts for deployment), deployment (release and distribute a version of a product to production).

Continuous Deployment (CD)—A process that validates software from a delivery staging environment and releases it to a production environment.

Continuous Flow—Smoothly moving software changes from the first step of a process to the last with minimal interruptions between steps.

Continuous Integration (CI)—A development practice that requires developers to merge their code into a common shared repository—ideally, multiple times per day.

Continuous Monitoring (CM)—This is a class of terms relevant to logging, notifications, alerts, displays, and analysis of test results information.

Continuous Security—See DevSecOps and Rugged DevOps.

Continuous Testing (CT)—This is a class of terms relevant to testing and verification of an EUT in a DevOps environment.

Culture—The values and behaviors that contribute to the unique social and psychological environment of an organization.

Deployment Pipeline—The flow of software changes into production via an automated software production line.

DevOps—The application of the lean principles of *Continuous Flow*, *Feedback*, and *Improvement* to people, process, and technology for increasing agility, stability, security, efficiency, quality, availability, and satisfaction.

DevOps Toolchain—A set of software tools that are linked (or chained) together to form a DevOps CI/CD pipeline.

DevSecOps—A mindset that "everyone is responsible for security" with the goal of safely distributing security decisions at speed and scale to those who hold the highest level of context without sacrificing the safety required.

Feature Toggle—The practice of using software switches to hide or activate features. This enables continuous integration and testing of a feature with selected stakeholders.

Function-as-a-Service (FaaS)—A category of cloud computing services that provides a platform allowing customers to develop, run, and manage application functionalities without the complexity of building and maintaining the infrastructure typically associated with developing and launching an app.

Governance—The establishment of policies, and continuous monitoring of their proper implementation, by the members of a governing body. It includes the mechanisms required to balance the powers of the members (with the associated accountability) and their primary duty of enhancing the prosperity and viability of the organization.

Hybrid Cloud—A cloud computing environment that uses a mix of on-premises, private cloud, and third-party public cloud services with orchestration between the two platforms.

Idempotent—The desired state of a server is defined as code or declarations, and the execution of configuration steps are automated to consistently achieve the defined server configuration state time after time. Configuration management tools offer idempotency CM tools (e.g., Puppet, Chef, Ansible, and SaltStack).

Immutable Infrastructures—A software deployment method in which changes to software modules on a deployment node (e.g., server, container, etc.) are replaced to ensure proper behavior, instead of instantiating an instance with error-prone, time-consuming patches and upgrades or mutations.

Improvement Kata—A structured way to create a culture of continuous learning and improvement.

Infrastructure-as-Code (IaC)—This is an approach for managing and provisioning computer data centers and cloud services through machine-readable definition files, rather than physical hardware configuration or interactive configuration tools. IaC approaches are promoted for cloud computing as infrastructure-as-a-service (IaaS).

Infrastructure-as-a-Service (IaaS)—On-demand access to a shared pool of configurable computing resources. Example service providers are AWS, Azure, GCS, IBM, Oracle.

IT Infrastructure Library (ITIL)—A set of best practice publications for IT service management published in a series of five core books representing the stages of the IT service life cycle, which are: Service Strategy, Service Design, Service Transition, Service Operation, and Continual Service Improvement.

Jidoka—Sometimes is called autonomation, this means automation with human intelligence. Jidoka highlights the causes of problems because work stops immediately when a problem first occurs. This leads to improvements in the processes that build in quality by eliminating the root causes of defects.

Kaizen—The practice of continuous improvement.

Kata—A lean management term refers to two linked behaviors: improvement kata and coaching kata. Improvement kata is a repeating a four-step routine by which an organization improves and adapts. It makes continuous improvement a daily habit through the scientific problem-solving method of plan, do, check, and act (PDCA).

Microservices—A software architecture that is composed of smaller modules that interact through APIs and can be updated without affecting the entire system. Microservices is a special case of an implementation approach for service-oriented architectures (SOA) used to build flexible, independently deployable software systems. Services in a microservice architecture are processes that communicate with each other over a network to fulfill a goal. The microservices approach is a first realization of SOA that followed the introduction of DevOps and is becoming more popular for building continuously deployed systems.

Muda—This is a Japanese term associated with lean manufacturing systems that means "waste."

Multi-Cloud—The use of multiple cloud computing and storage services in a single heterogeneous architecture. For example, an enterprise may concurrently use separate cloud providers for infrastructure (IaaS) and software (SaaS) services or use multiple infrastructure (IaaS) providers.

Operations (Ops)—The individuals or team involved in the daily operational activities of IT systems and services.

Orchestration—Tasks, usually automated, that setup an environment for a system to operate. Alternative: An approach to building automation that interfaces or "orchestrates" multiple tools together to form a toolchain.

Platform-as-a-Service (PaaS)—Category of cloud computing services that provides a platform allowing customers to develop, run, and manage applications without the complexity of building and maintaining the infrastructure

Service Catalog—Subset of the Service Portfolio that consists of services that are live or available for deployment. It has two aspects: the Business/Customer Service Catalog (visible to customers) and the Technical/Supporting Service Catalog.

Site Reliability Engineering (SRE)—This is a discipline that incorporates aspects of software engineering and applies them to infrastructure and operations problems. The main goals are to create ultra-scalable and highly reliable software systems.

Software Version Management System—A repository tool which is used to manage software changes. Examples are Git, GitHub, GitLab, and Perforce.

Software-as-a-Service (SaaS)—A category of cloud computing services in which software is licensed on a subscription basis.

The Three Ways—Maturity levels of DevOps described in *The Phoenix Project—Continuous Flow, Continuous Feedback,* and *Continuous improvement.*

Toolchain—A set of distinct software development tools that are linked (or chained) together by specific stages to automate an end-to-end CI/CD pipeline.

Trunk—The primary source code integration repository for a software product.

User Acceptance Testing (UAT)—End-users testing from the point of view of usability.

Value Stream—All the activities to go from a customer request to a delivered product or service.

Value-Stream Management—Mapping, optimizing, visualizing, and governing business value flow (including epics, stories, and work items) through heterogeneous enterprise software delivery pipelines to operations.

Value-Stream Mapping—A visualization that depicts the flow of information, materials, and work across functional silos with an emphasis on identifying and quantifying waste, including time and quality.

Version Control Repository—A repository where developers can commit and collaborate on their code. It also tracks historical versions and potentially identifies conflicting versions of the same code.

Version Control Tools—Same as software version management tool.

Yokoten—A Japanese term used in lean manufacturing referring to the sharing of recommended engineering practices.

Appendix B

DevOps Transformation Application Scorecard

DevOps Transformation Application Scorecard

Name of Application :

Some ***applications*** can benefit from DevOps more readily than others. An average rating score of 3 or more is preferred.

Rating scores: 0 = "Don't know/unsure", 1 = "Doesn't fit this characteristic", 2 = "Somewhat fits", 3 = "Mostly fits", 4 = "Good fit", 5 = "Perfect fit"	**Rating (0, 1 to 5)**
1. **Lead time**: The ***application*** will benefit from faster lead times, where lead time = time from backlog to deployment.	
2. **Leadership**: Leaders over this ***application*** are open to collaboration and will be sponsors of change.	
3. **Culture**: Team players that are associate with the ***application*** team (Product owners, Dev, QA, Ops, Infra, Sec, PM) are open to collaboration and change.	
4. **Application architecture**: The ***application*** is currently using, or planning to use, service-oriented, modular architectures.	
5. **Product Team size**: At least 15 people associated with the ***application*** team (includes Product owners, Dev, QA, Ops, Infra, Sec, PM) . Smaller projects may not show strong impact or justify substantive DevOps investment.	
6. **Duration**: The ***application*** is expected to be undergoing changes for more than a year. yield ROI.	
7. **Impact/Risk**: The ***application*** represents a good level of business impact and visibility but does not involve an extreme amount of risk for the business.	
8. **Frequent changes**: The ***application*** experiences frequent demands for changes from the business.	
9. **Tools**: The tools in the ***application*** toolchain will not all need to be replaced to implement a DevOps toolchain.	
10. **Effort per release**: Efforts to build, test and/or deploy releases of the ***application*** is significant and could be reduced significantly by automaton.	
Average	

An Excel version of this scorecard can be found on www.EngineeringDevOps.com

Appendix C

DevOps Transformation Vision Meeting

Example presentation. The template is available in PowerPoint format on:

www.engineeringDevOps.com

DevOps Transformation Visioning Meeting

Attendees
DevOps Sponsor
DevOps Initiator
DevOps Expert

Strategic Goals for the DevOps Transformation

At this level the goals are expressed as visionary and qualifiable, but not necessary quantifiable objectives which will have a major impact on the purpose of the organization. These are not technical goals as much as they are business goals. For example, statements about improvements to competitiveness, innovation, customer satisfaction, employee satisfaction would qualify as visionary strategic goals. Technologies expected to accomplish these goals may be included but the emphasis of these goals are on strategic outcomes instead of the means to accomplish them. Usually there are no more than one or two strategic goals because the intent is to have a simply stated, clear, common visionary goal that the entire organization can understand and rally around. Strategic goals do not substitute for tactical goals. It is important to explain WHY this goal was chosen over alternatives considered. An example strategic goal is shown below:

Our product organization shall be the market leader, measured by volume of sales, within three years, through increasing new product release velocity ten times while reducing non-value added costs in the delivery pipeline two times.

WHY? The market for this product category is maturing. Only the top three competitors are expected to survive the next 5 years. Industry trends show the winning companies will the ones with the fastest, most innovative value-stream and lowest price.

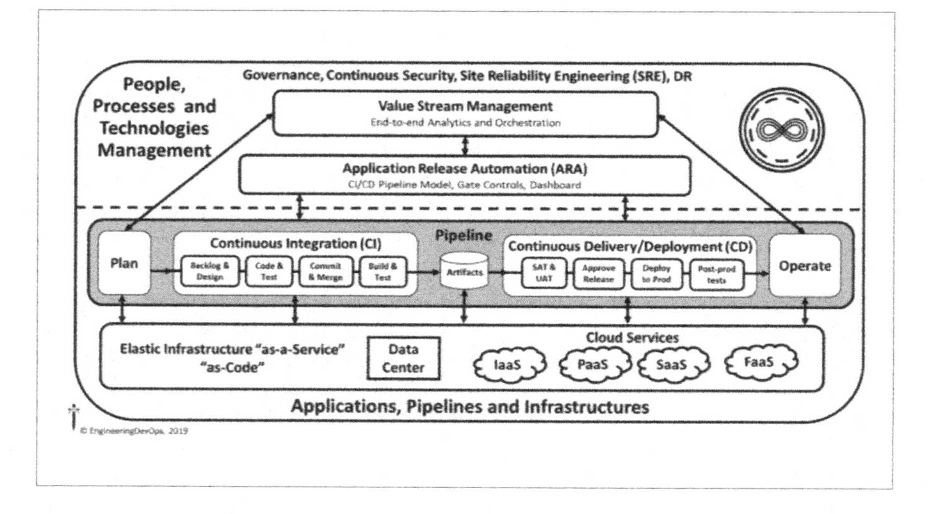

DevOps Transformation Application Selection

The DevOps transformation will apply to the following applications: <name of application(s)>

The <name> application will be used as a model for kicking-off and evaluating changes needed to people, process and technology for the DevOps transformation.

WHY? A DevOps Application Selection Scorecard indicates these application(s) will most likely benefit from the DevOps transformation.

DevOps Seven-Step Transformation Engineering Blueprint

1. Visioning
Strategic, Sponsors, Partners
2 Alignment
Leadership, Team, Applications
3. Assessment
Discovery, 9 Pillars Maturity, Deep-Dive, Value-Stream Map
4. Solution
Future State, Road-map, Epics, Re-alignment

5. Realize
Projects, Stories, Tasks, POC, Validation, Training, Deployment, Governance
6. Operationalize
Monitor SLI/SLO/SLA, SRE Controls, Retrospectives
7. Expansion
Continuous Flow, Enterprise Adoption, Continuous Feedback, Continuous Improvement

DevOps Transformation Team Leaders

Name	Role/Responsibility
	DevOps Transformation Sponsor
	DevOps Expert
	Product Development
	Product Quality Assurance
	Operations Support
	Infrastructure
	DevOps Tools
	Partner Management
	Product Owner
	Security
	Project Management
	Training
	Finance
	Human Resources
	Governance

Identify strategic level leaders and partners that cover the cross-section of organization functions that are most relevant to the DevOps transformation. For example, strategic leaders with responsibility over roles from Development, Quality Assurance, Operations, Infrastructure, DevOps Tools, 3rd party suppliers, Product Owners, Security, Project Management, Training, Finance, Human Resources and Governance should be considered, depending on the extent to which each of the roles are likely to be affected by the transformation.

Appendix D

DevOps Transformation Goals Scorecard

The template on the following page is available in Excel format on:

www.engineeringDevOps.com

DevOps Transformation Goals Scorecard	Metric Unit	Importance (I) (1-5) (1=low, 5=critical)	Current State	Desired State	Percent Improvement (P%)	SCORE I x P%	RANK
Agility		4.4			205%	9	1
Lead time: Duration from code commit until code is ready to be deployed to production.	# days	5	5.0	2.0	250%	13	2
Release Cadence: Frequency of having releases ready for deployment to live production.	# releases / month	5	0.2	0.6	300%	15	1
Fraction of Non-Value Added-Time Fraction of their time employees are not spending on new value enhancing work such as new features or code.	Fraction %	3	30%	20%	150%	5	15
Batch size: Product teams break work into small batch increments.	1-10 (1 rarely, 10 usually)	5	3.0	5.0	167%	8	3
Visible Work: Workflow is visible throughout the pipeline.	1-10 (1 rarely, 10 usually)	4	5.0	8.0	160%	6	6
Security		2.7			233%	6	2
Security Events: # times that a serious business impacting security event occur over a set period	# per year	2	2	1	400%	8	4
Unauthorized Access: # times per period that unauthorized users accessed unathorized information.	# per year	3	1	1	100%	3	21
Fraction of time remediating security problems: Average % of time that employees spend remediating security issues.	%	3	5%	3%	200%	6	9
Satisfaction		3.8			137%	5	5
Employee satisfaction with team: Employees are likely to recommend their team as a great to work with.	1-10 (1 not likely, 10 most likely)	3	7.0	8.0	114%	3	20
Employee satisfaction with organization: Employees are likey to recommend their organization as a organization to work in.	1-10 (1 rarely, 10 most likely)	4	6.0	8.0	133%	5	13
Organization type: The culture is of the organization is a *generative* type, with good communication flow, cooperation and trustful.	1-10 (1 rarely, 10 very much)	4	5.0	7.0	140%	6	11
Leader Style for Recognition: Leaders promote personal recognition by commending team for better-than-average work, acknowledging improvement in quality of work and personally compliments individuals' outstanding work.	1-10 (1 rarely, 10 very much)	4	5.0	8.0	160%	6	6
Stability		4.0			152%	6	3
MTTR: Mean-Time-To-Recover (MTTR) from failure/service outage in production.	hours	4	1.0	0.7	154%	6	8
Code merges problems: % of code merges from development branches to the trunk branch break the trunk branch.	%	4	15%	10%	150%	6	10
Quality		3.7			148%	5	4
Failures in Production: Frequently of failures requiring immediate remediation occur in live production.	# per week	4	0.1	0.05	200%	8	4
Test and Data Available: Tests and test data are sufficient and readily available when needed.	1-10 (1 rarely, 10 usually)	3	7.0	9.0	129%	4	19
Customer Feedback: The organization regularily seeks customer feedback and incorporates the feedback into design.	1-10 (1 rarely, 10 usually)	4	7.0	8.0	114%	5	14
Efficiency		3.3			143%	5	6
Unplanned work: % of time do employees spend on all types of unplanned work, including rework.	%	3	15%	10%	150%	5	15
Operating Costs are Visible: Comprehensive metrics are kept for *operating* costs of development and operations.	1-10 (1 rarely, 10 usually)	3	5.0	7.0	140%	4	17
Capital costs are visable: Comprehensive metrics are kept for *capital* costs of development and operations.	1-10 (1 rarely, 10 usually)	3	5.0	7.0	140%	4	17
Backlog Visibility: Lean product management is practiced using highly visible, easy-to-understand presentation formats that show work to be done.	1-10 (1 rarely, 10 usually)	4	5.0	7.0	140%	6	11

The DevOps Transformation Goal Scorecard spreadsheet is posted on www.EngineeirngDevOps.com

Appendix E

DevOps Transformation Alignment Meeting

Example presentation. The template is available in PowerPoint format on:

www.engineeringDevOps.com

DevOps Transformation
Alignment Meeting

Attendees
DevOps Sponsor
DevOps Initiator
DevOps Expert

DevOps Transformation Team Leaders

Name	Role/Responsibility
	DevOps Transformation Sponsor
	DevOps Expert
	Product Development
	Product Quality Assurance
	Operations Support
	Infrastructure
	DevOps Tools
	Partner Management
	Product Owner
	Security
	Project Management
	Training
	Finance
	Human Resources
	Governance

Identify strategic level leaders and partners that cover the cross-section of organization functions that are most relevant to the DevOps transformation. For example, strategic leaders with responsibility over roles from Development, Quality Assurance, Operations, Infrastructure, DevOps Tools, 3rd party suppliers, Product Owners, Security, Project Management, Training, Finance, Human Resources and Governance should be considered, depending on the extent to which each of the roles are likely to be affected by the transformation.

Strategic Goals for the DevOps Transformation

Our product organization shall be the market leader, measured by volume of sales, within three years, through increasing new product release velocity ten times while reducing non-value added costs in the delivery pipeline two times.

WHY? The market for this product category is maturing. Only the top three competitors are expected to survive the next 5 years. Industry trends show the winning companies will the ones with the fastest, most innovative value-stream and lowest price.

At this level the goals are expressed as visionary and qualifiable, but not necessary quantifiable objectives which will have a major impact on the purpose of the organization. These are not technical goals as much as they are business goals. For example, statements about improvements to competitiveness, innovation, customer satisfaction, employee satisfaction would qualify as visionary strategic goals. Technologies expected to accomplish these goals may be included but the emphasis of these goals are on strategic outcomes instead of the means to accomplish them. Usually there are no more than one or two strategic goals because the intent is to have a simply stated, clear, common visionary goal that the entire organization can understand and rally around. Strategic goals do not substitute for tactical goals. It is important to explain WHY this goal was chosen over alternatives considered. An example strategic goal is shown above.

Definition of DevOps

DevOps is the application of lean practices Continuous Flow, Continuous Feedback and Continuous Improvement to people, process and technology for the benefit of agility, stability, efficiency, quality, security and satisfaction.

DevOps_The_Gray

There is no industry standard definition of DevOps yet it is important for each organization to have a definition the team can accept and rally around. The above definition is suggested. Alternative definitions are also acceptable provided the DevOps Transformation Team accepts a definition for use in the organization.

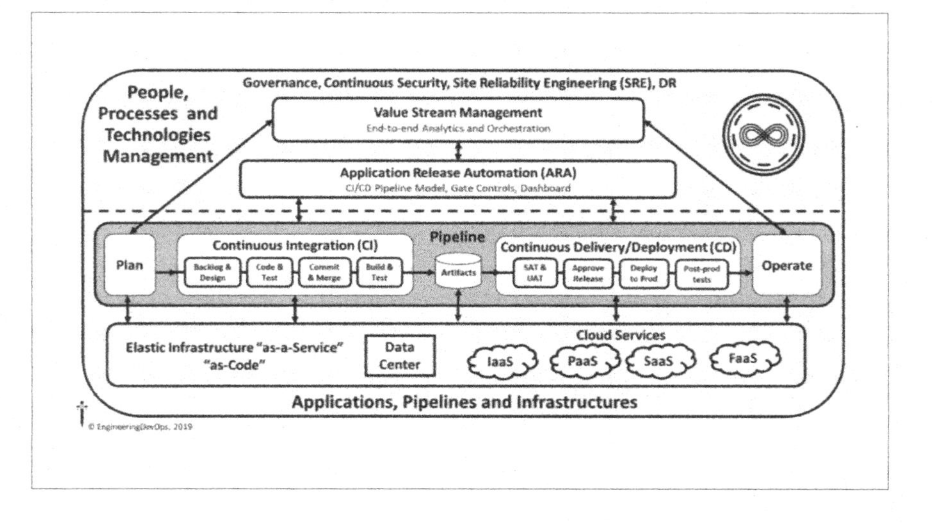

DevOps Transformation Application Selection

The DevOps transformation will apply to the following applications: <name of application(s)>

The <name> application will be used as a model for kicking-off and evaluating changes needed to people, process and technology for the DevOps transformation.

WHY? A DevOps Application Selection Scorecard indicates these application(s) will most likely benefit from the DevOps transformation.

DevOps Maturity Levels

What is the Maturity Level of the Selected Application?

Maturity Level	People	Process	Technology
Chaos	• Silo teams and organization with little communication between silos • Blame and finger-pointing • Dependent on experts	• Requirements, planning and tracking processes poorly defined and operated manually • Unpredictable and reactive	• Manual builds and deployments • Manual quality assurance • Environment inconsistencies
Continuous Integration	• Managed communications between silos • Limited knowledge sharing • Ad hoc training	• Processes defined within silos • No standards for end-to-end processes • Can repeat what is known but can't react to unknowns	• Source code version management • Automated builds, release artifacts, & automated tests • Painful but repeatable releases
Continuous Flow ("First Way" of DevOps)	• DevOps leadership • Collaboration between cross-functional teams • DevOps training program	• End-to-end pipeline automated • Standards across the org for applications, releases processes and infrastructure	• Toolchain orchestrates and automates builds, tests and packaging deliverables. • Infrastructure is orchestrated as code. • Automated metrics and analysis for release acceptance and deployment.
Continuous Feedback ("2nd Way" of DevOps)	• Collaboration based on shared metrics with a focus on removing bottlenecks • SLIs, SLOs and SLAs • DevOps Mentors and Guilds	• Pro-active monitoring • Metrics collected and analyzed against business goals • Visibility and repeatability	• Applications, pipelines and infrastructure fully instrumented • Metrics and analytics dashboards • Orchestrated deployments with automated rollbacks
Continuous Improvement ("3rd Way" of DevOps)	• Culture of continuous experimentation and improvement	• Self service automation • Risk and cost optimization • High degree of experimentation	• Zero downtime deployments • Immutable infrastructure • Actively enforce resiliency by forcing failures

DevOps Seven-Step Transformation Engineering Blueprint

1. Visioning
Strategic, Sponsors, Partners
2 Alignment
Leadership, Team, Applications
3. Assessment
Discovery, 9 Pillars Maturity, Deep-Dive, Value-Stream Map
4. Solution
Future State, Road-map, Epics, Re-alignment

5. Realize
Projects, Stories, Tasks, POC, Validation, Training, Deployment, Governance
6. Operationalize
Monitor SLI/SLO/SLA, SRE Controls, Retrospectives
7. Expansion
Continuous Flow, Enterprise Adoption, Continuous Feedback, Continuous Improvement

DevOps Transformation Goals Workshop

For each goal enter Unit, Importance (I), Current State, Desired State.

Once all goal data is entered calculate % Improvement, Score and Rank to determine the top goals.

Enter | Calculate

DevOps Transformation Goals Scorecard

20 goals in six categories: Agility, Stability, Efficiency, Quality, Security, and Satisfaction

DevOps Transformation Practices Score Workshop

For each Practice Category enter Importance and Current Level of Practice Score

Once all data is entered calculate a score and rank to determine the top priority practices.

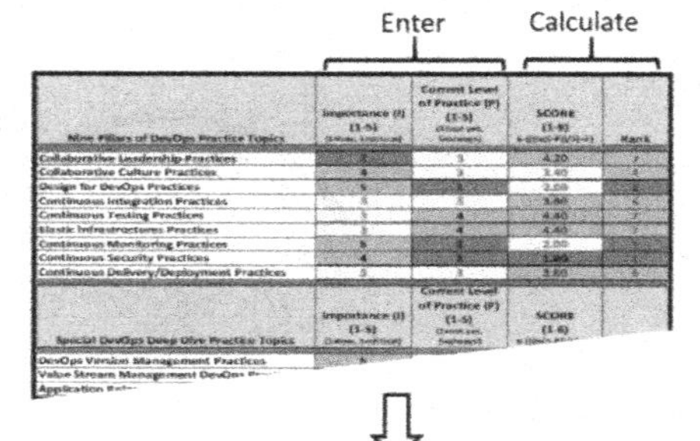

Nine Pillars and 11 Deep-Dive Topics

DevOps Transformation Assessment Workshops Planned Agenda

- Recap DevOps Goals
- Summary of Discovery Surveys
- Maturity Assessment Workshops
- Current State Value Stream Map Workshop

Appendix F

DevOps Transformation Practices Topics Scorecard

The DevOps Transformation Practices Topics Scorecard on the following page is posted in spread-sheet form on:

www.engineeringDevOps.com

DevOps Transformation Practices Topics

Nine Pillars of DevOps Practice Topics	Importance (I) (1-5) (1=low, 5=critical)	Current Level of Practice (P) (1-5) (1=not yet, 5=always)
Collaborative Leadership Practices		
Collaborative Culture Practices		
Design for DevOps Practices		
Continuous Integration Practices		
Continuous Testing Practices		
Elastic Infrastructures Practices		
Continuous Monitoring Practices		
Continuous Security Practices		
Continuous Delivery/Deployment Practices		
Special DevOps Deep Dive Practice Topics	**Importance (I) (1-5) (1=low, 5=critical)**	**Current Level of Practice (P) (1-5) (1=not yet, 5=always)**
DevOps Version Management Practices		
Value Stream Management DevOps Practices		
Application Release Automation Practices		
DevOps Infrastructure-As-Code		
Hybrid Cloud DevOps Practices		
Multi-Cloud DevOps Practices		
Application Performance Monitoring Practices		
DevOps Training Practices		
Site Reliability Engineering Practices		
DevOps Service Catalog		
DevOps Governance Practices		
Other >		

The DevOps Transformation Practices Topics Scorecard is posted in spreadsheet form on www.EngineeringDevOps.com

Appendix G

DevOps Assessment Discovery Survey

This DevOps Assessment Discovery Survey requests information regarding the **current state** of a specific application or group of applications that have similar current state information. The answers should NOT be a projection of future state.

This survey requests information about the following:

A) **Application** that is being assessed
B) **Organization** that applies to the Application being assessed
C) **Pipeline** that applies to the Application being assessed
D) **Tools** that are used to support the pipeline
E) **Infrastructure** that applies the Application being assessed
F) **Commercial-Off-The-Shelf (COTS) systems** that applies the Application being assessed

A) Application Assessment Discovery

NAME OF APPLICATION: ______________________

1. What is the primary purpose of the Application from the point of view of key customers and other key stakeholders?
2. Identify and characterize the customers and major stakeholders of the Application (External and Internal).

3. What is the value that customers expect to derive from the Application?
4. Comment on the average number of defects reported by customers for this Application. Is the number acceptable? Is the number increasing or decreasing?
5. What do customers most like about the Application?
6. What do customers least like about the Application?
7. Provide a high-level block level architecture of the Application showing major components and relationships between components. Please Include monolithic components, services, microservices, three-tier app, database elements, user interfaces, third-party components, COTS components, etc.
8. Describe the internal construction of each of the major components used to construct the Application (Software model: languages, service model, data model, etc.).
9. What are the strengths and weaknesses of the Application from an architecture point of view?
10. What are the major future roadmap items planned for your Application?

B) Organization Assessment Discovery

1. Provide a schematic organization chart for the Application that shows relationships between Leaders, Developers, QA, Ops staff, Infrastructure, tools teams, Security, Product Owners, Product Management, and Partner Management. This does not need to be a detailed org chart. It is the reporting structure and general organization structure (layers and reporting lines between functions) that is important to identify. Include an estimate of the effective number of people for each function on the org chart and their geographic locations.
2. Where in the organization does responsibility for selection of Dev, QA, and Ops tools reside—in separate departments, or in a common department that supports tools for the entire organization?

3. To what extent do Development, QA, Ops leaders, and team members collaborate to accomplish product changes and production deployments?
4. To what extent are Dev, QA, and Ops in agreement regarding DevOps goals?

C) Pipeline Assessment Discovery

1. What at the stages in your pipeline called (e.g., Dev, Integration, Staging, Deployment, etc.)?
2. How long does it take, on average, to do a successful complete build of the software for the entire application (including unit tests and smoke tests)?
3. What is the longest time that a successful incremental build takes for the Application (including unit tests and smoke tests)?
4. Are feature branches managed in separate software version management systems to control software changes, or are all software changes made from a common trunk branch managed by a common software version management system?
5. How often do developers check in code to the trunk (sometimes called the release) branch?
6. What percentage of developers' code check-ins pass through trunk integration without any problems? (Or the opposite: what percentage of code check-ins have problems that "break" the trunk build?)
7. What percentage of smoke tests are automated?
8. What percentage of regression tests are automated?
9. What percentage of release acceptance tests are automated?
10. Are containers used to package and deploy code changes?
11. How are release decisions made? What are the criteria?
12. After a release decision is made, how long does it take to release to production?
13. After release to production, how long does it take to complete deployment to all deployment nodes?

14. Identify whether Dark Launch strategies are used for deployment (e.g., Green/Blue, Rolling A/B, Rolling Canary, etc.).

D)Tools

Identify tools are used currently for each category listed in this section. Include the name of any homegrown/DIY, open-source, or commercial tools that are currently being used. Flag any tools that are in use but having problems or are not popular. If there is another tool not currently being used but being considered, identify it as such and give the reasons why.

1. Communication between teams (e.g., ChatOps, Splunk, email, etc.)
2. Software IDEs (e.g., Visual Studio, etc.)
3. Software version management (e.g., GitHub, GitLab, Perforce Bitbucket, etc.)
4. Code collaboration (e.g., Gerrit, Code Collaborator, etc.)
5. Unit test tools (e.g., Junit, Mocks, Gradle, etc.)
6. Build automation (e.g., Ant, Maven, Jenkins, CloudBees, Bamboo, Team City, Cruise Control, etc.)
7. Static code analysis (e.g., SonarQube, Coverity, etc.)
8. Test creation tools (e.g., Tricentis, Cucumber, etc.)
9. API test tools (e.g., SoapUI, etc.)
10. User interface test automation (e.g., Selenium, TestPlant, Ranorex, etc.)
11. Test management (e.g., HP Quality Center etc.)
12. Load and performance testing (e.g., HP LoadRunner, JMeter, etc.)
13. Test environment orchestration (e.g., Quali, Spirent, etc.)
14. Application Release Automation (ARA) (e.g., Electric Cloud, XebiaLabs, etc.)
15. CI/CD pipeline tools (e.g., Codeship, Travis, Circle CI, etc.)
16. Image repository (e.g., JFrog, Nexus, Archiva, etc.)
17. Deployment (e.g., Harness, Electric Cloud, Urban Code, etc.)
18. Application Performance Monitoring (APM) (e.g., New Relic, App Dynamics, Dynatrace, etc.)

19. Infrastructure monitoring tools (e.g., Splunk, Xenoss, Nagios, etc.)
20. Security scanning and monitoring tools (e.g., Appscan, Twistlock, Nexpose, Fortify, Veracode, etc.)
21. Security secrets management tools (e.g., HashiCorp Vault, etc.)
22. A/B test tools
23. Green/Blue deployment tools
24. Canary test tools
25. Configuration management tools (e.g., Puppet, Chef, Ansible, SaltStack, etc.)
26. Infrastructure as Code tools (e.g., Terraform, etc.):
27. Virtual machines (e.g., VMware, etc.)
28. Cloud services (e.g., AWS, Azure, GCS, IBM Bluemix, etc.)
29. Containers and containers orchestration (e.g., Docker, Rancher, Kubernetes, OpenShift, etc.)
30. Database tools (e.g., Oracle, MSQL, etc.)
31. Database pipeline tools (e.g., DB Maestro, Datical, etc.)
32. Data management tools (e.g., Delphix. etc.)
33. Database monitoring tools (e.g., Visual Cortex, etc.)

E) Infrastructure

1. Describe the deployment infrastructure that is in use for this application. Include network, compute and storage elements. Identify regional differences. Include a diagram.
2. Identify IaaS, PaaS, and SaaS infrastructure in use.
3. What are the strengths of the current infrastructure?
4. What are the weaknesses of the current infrastructure?
5. How is governance of the infrastructure managed?
6. How are infrastructure change requests handled?
7. What is the average utilization of the infrastructure?
8. How is high availability assured by the infrastructure?
9. How is disaster recovery handled for the infrastructure?
10. How is resilience of the infrastructure assured (e.g., Chaos monkey, etc.)?

F) Commercial Off-The-Shelf (COTS) systems

For each COTS system in use, identify the following items:

1. Name of the COTS system
2. Version in use
3. Frequency of changes (e.g., new releases or updates)
4. How often are releases of the model application affected by COTS systems failures?
5. How are changes of the COTs systems tested with the application before they are used in production?
6. Describe the process for deploying new versions of the COTs system.
7. How long does it take to deploy a new version of the COTs system?
8. What systems are used to monitor the performance of the COTS system?

Appendix H

DevOps Transformation Practices Maturity Assessment Workshop

This appendix shows an outline for the DevOps Transformation Practices Maturity Assessment Workshop. There are too many practices to list in this document, and only a subset of them are relevant for any specific Application assessment.

The DevOps Transformation Practices Maturity Assessment Scorecards with a complete, up-to-date set of practices to choose from for a specific application assessment is posted in spreadsheet form on:

www.engineeringDevOps.com

DevOps Transformation Practices Maturity Assessment Workshop

DevOps Transformation Practices Maturity Assessment Workshop

This workshop scores practices for :

- DevOps 9 pillars
- Deep-Dive Topics

Scope:
Application selected by the DevOps Transformation Leadership Team
Importance Score is relative to the DevOps Transformation Goal
*Practice Level is the **Current** practice level.*

The calculated Scores are used to determine the Rank for practices of the 9 Pillars and Deep-Dive topics that are most important to improve to meet the DevOps Transformation goals.

DevOps 9 Pillars Practices Maturity Workshop

For each practice enter Importance (I) for meeting the DevOps Transformation Goal and the **Current** Level of Practice (P)

Once all data is entered calculate Score and GAP Rank to determine the highest gapped practices.

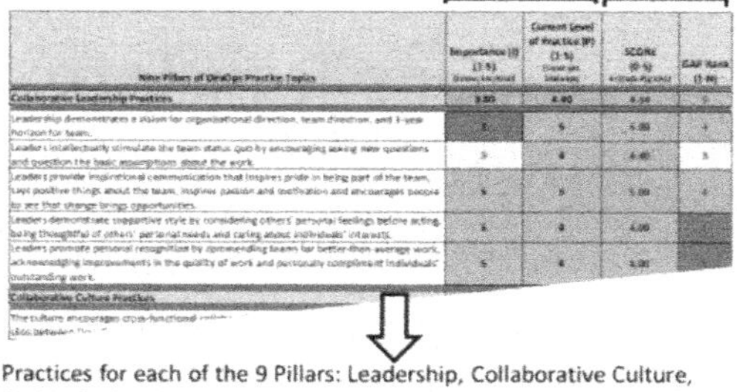

Enter

Calculate

Practices for each of the 9 Pillars: Leadership, Collaborative Culture, Design for DevOps, Continuous Integration, Continuous Testing, Elastic Infrastructure, Continuous Monitoring, Continuous Security and Continuous Delivery/Deployment

DevOps 9 Pillars Practices Maturity Workshop

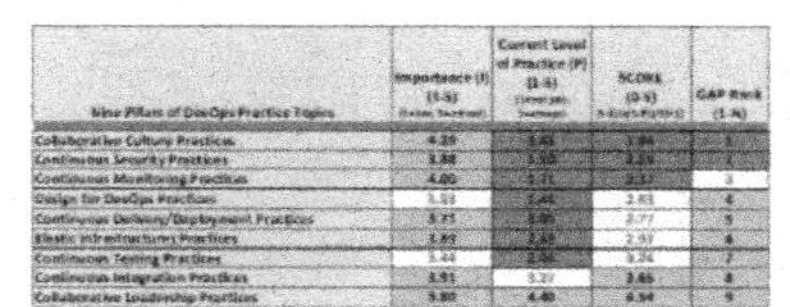

Nine Pillars of DevOps Practice Topics	Importance (I) (1-5)	Current Level of Practice (P) (1-5)	SCORE (0-5)	GAP Rank (1-N)
Collaborative Culture Practices	[illegible]	[illegible]	[illegible]	1
Continuous Security Practices	3.88	[illegible]	[illegible]	2
Continuous Monitoring Practices	4.00	[illegible]	[illegible]	3
Design for DevOps Practices	[illegible]	[illegible]	[illegible]	4
Continuous Delivery/Deployment Practices	[illegible]	[illegible]	[illegible]	5
Elastic Infrastructure Practices	[illegible]	[illegible]	[illegible]	6
Continuous Testing Practices	[illegible]	[illegible]	[illegible]	7
Continuous Integration Practices	3.91	[illegible]	[illegible]	8
Collaborative Leadership Practices	[illegible]	4.40	[illegible]	9

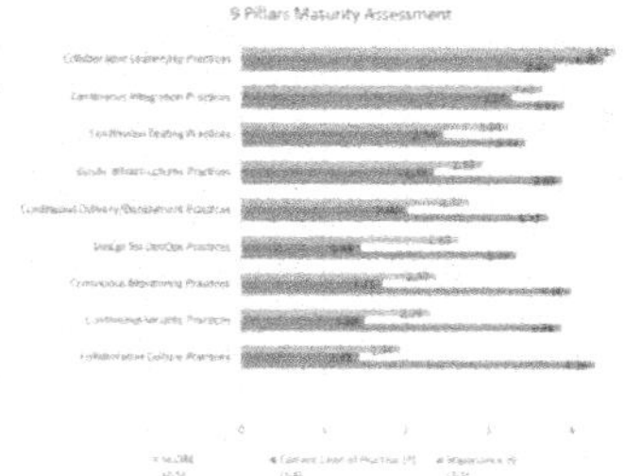

The calculated Scores are used to determine the Rank indicated which Pillar and practices are most important to improve to meet the DevOps Transformation goals.

DevOps Deep-Dive Topics Practices Maturity Workshop

Enter Calculate

For each practice enter Importance (I) for meeting the DevOps Transformation Goal and the **Current** Level of Practice (P)

Once all data is entered calculate Score and GAP Rank to determine the highest gapped practices.

Practices for each of the Deep-Dive topics that are selected for assessment: Training Practices, Governance Practices, Value-Stream Management Practices, Application Performance Monitoring (APM) Practices, Site Reliability Engineering (SRE) Practices, Service Catalog Practices, Application Release Automation (ARA) Practices, Multi-Cloud Practices, Infrastructure-as-Code (IaC) Practices, Hybrid Cloud Practices, Version Management Practices

DevOps Deep-Dive Topics Practices Maturity Workshop

DevOps Topics Maturity Assessment Scorecard	Importance (I) (1-5)	Current Level of Practice (P) (0-5)	SCORE (0-5)	GAP Rank (1-N)
DevOps Training Practices	3.86	[illegible]	[illegible]	[illegible]
DevOps Governance Practices	4.11	[illegible]	[illegible]	[illegible]
DevOps Value Stream Management Practices	[illegible]	[illegible]	[illegible]	3
DevOps Application Performance Monitoring (APM) Practices	3.87	[illegible]	[illegible]	4
DevOps Site Reliability Engineering (SRE) Practices	3.75	[illegible]	[illegible]	5
DevOps Service Catalog Practices	3.75	[illegible]	2.56	6
DevOps Application Release Automation (ARA) Practices	3.75	[illegible]	[illegible]	7
DevOps Multi-Cloud Practices	3.92	[illegible]	[illegible]	8
DevOps Infrastructure as Code (IaC) Practices	4.00	[illegible]	3.53	9
DevOps Hybrid Cloud Practices	[illegible]	[illegible]	3.71	10
DevOps Version Management Practices	4.07	[illegible]	4.42	11

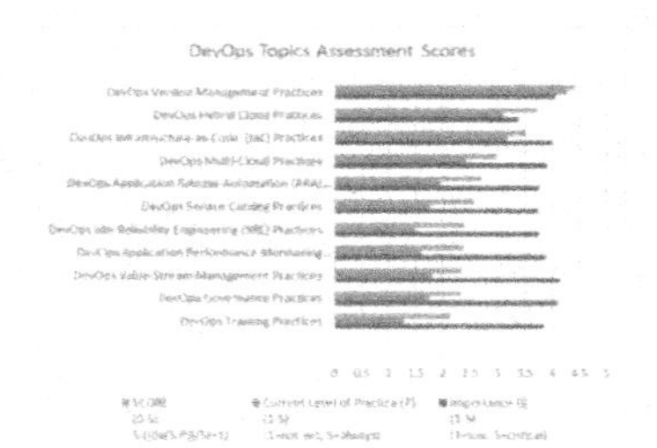

The calculated Scores are used to determine the Rank indicated which Deep-Dive topic and practices are most important to improve to meet the DevOps Transformation goals.

Appendix I

DevOps Current State Value-Stream Mapping Workshop

This appendix shows an outline for the DevOps Current State Value-Stream Mapping Workshop.

The DevOps Current State Value-Stream Mapping Workshop template is posted in PowerPoint form on:

www.engineeringDevOps.com

DevOps Engineering Value Stream Mapping Workshop (Current-State)

DevOps Engineering Value Stream Mapping Workshop (Current State)

This interactive workshop creates a **current state value-stream map** for the Application.

The value selected for analysis in this workshop is determined by the DevOps Leadership Team .

Factors affecting the selection of value stream:

- *Application selected by the DevOps Transformation Leadership Team*
- *DevOps Goal priorities*
- *DevOps Practices Gap Rank*

The output of this workshop is a key input to the Solution stage of the DevOps Transformation Blueprint.

DevOps Engineering Value Stream Mapping Workshop Step 1 – Select the Value and the Measurements

The DevOps Transformation Leadership team selects the Value or Values to be analyzed by Value-Stream Analysis, and the measurement to be used for the analysis of the selected Value Streams for the Application

Example:

- Value 1: Lead time: measured as the duration from the time a task is assigned in the backlog until the code is delivered, assuming not errors that require remediation.
- Value 2: Quality: measured as the % of changes integrated into trunk that do not require rework at the deployment stage.

DevOps Engineering Value Stream Mapping Workshop Step 2: Determine the Pipeline Stages relevant to the Values Selected for Analysis

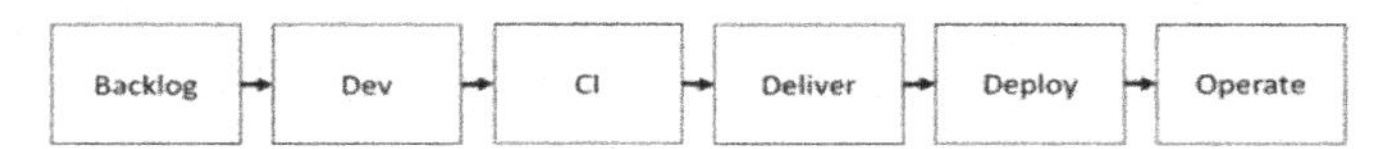

To qualify as a stage:

- Activities is relevant to the affecting the value being analyzed
- Quantifiable input criterion can be defined
- Quantifiable output criterion can be defined

DevOps Engineering Value Stream Mapping Workshop Step 3: Determine the measurements that will be used.

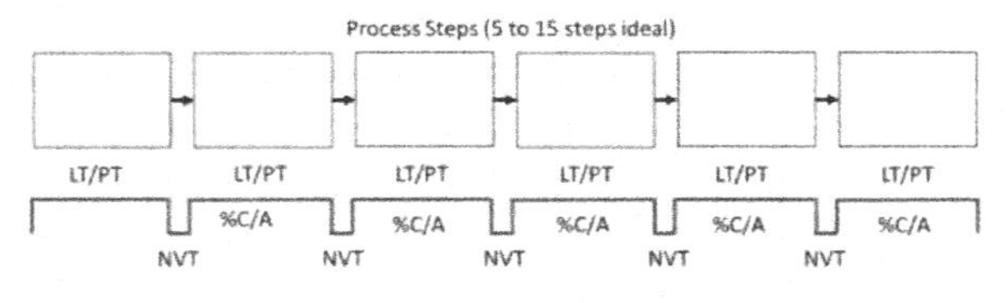

LT/PT: Value Added Time (Lead Time/Process Time)
NVT: Non-Value Added Time (e.g. Wait time)
%C/A: % of time process can use results of prior stage without rework

Example measurements:

- Lead Time and Process time for each stage
- NVT wait time between stages
- % of time next stage can use inputs without requesting rework

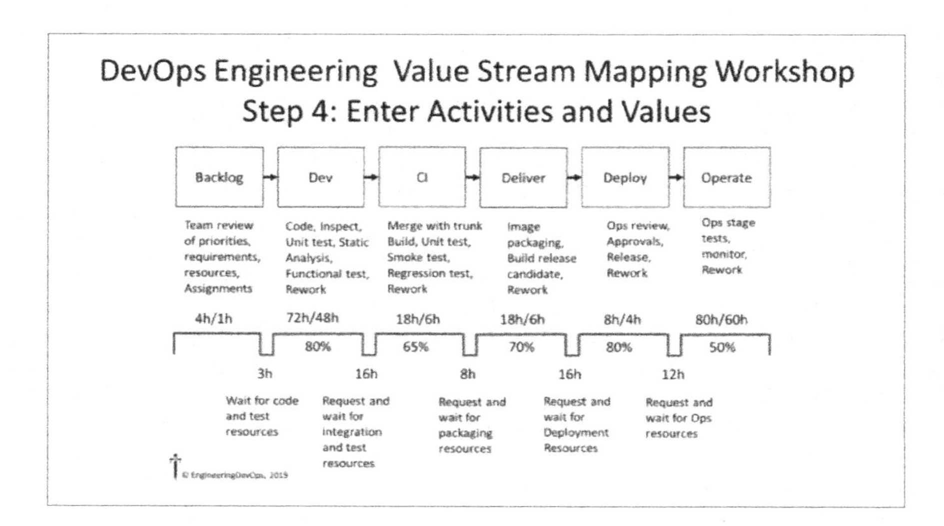
DevOps Engineering Value Stream Mapping Workshop
Step 4: Enter Activities and Values
Backlog
Dev
CI
Deliver
Deploy
Operate
Team review of priorities, requirements, resources, Assignments
Code, Inspect, Unit test, Static Analysis, Functional test, Rework
Merge with trunk Build, Unit test, Smoke test, Regression test, Rework
Image packaging, Build release candidate, Rework
Ops review, Approvals, Release, Rework
Ops stage tests, monitor, Rework
4h/1h
72h/48h
18h/6h
18h/6h
8h/4h
80h/60h
80%
65%
70%
80%
50%
3h
16h
8h
16h
12h
Wait for code and test resources
Request and wait for integration and test resources
Request and wait for packaging resources
Request and wait for Deployment Resources
Request and wait for Ops resources
© EngineeringDevOps, 2019

Appendix J

DevOps Solution Requirements Alignment Matrix

The template on the following page is available in Excel format on:

www.engineeringDevOps.com

DevOps Solution Requirements Alignment Matrix

High-Gap DevOps Practices	DevOps Sponsor	DevOps Expert	Product Development	Product Quality	Operations Support	Infrastructure	DevOps Tools	Partner Management	Product Owner	Security	Project Management	Training	Finance	Human Resources	Governance	Votes	Group Rank

complete, up-to-date set of practices, is posted in spreadsheet form on www.EngineeringDevOps.com

Appendix K

Value-Stream Map Template

The template on the following page is available in Excel format on:

www.engineeringDevOps.com

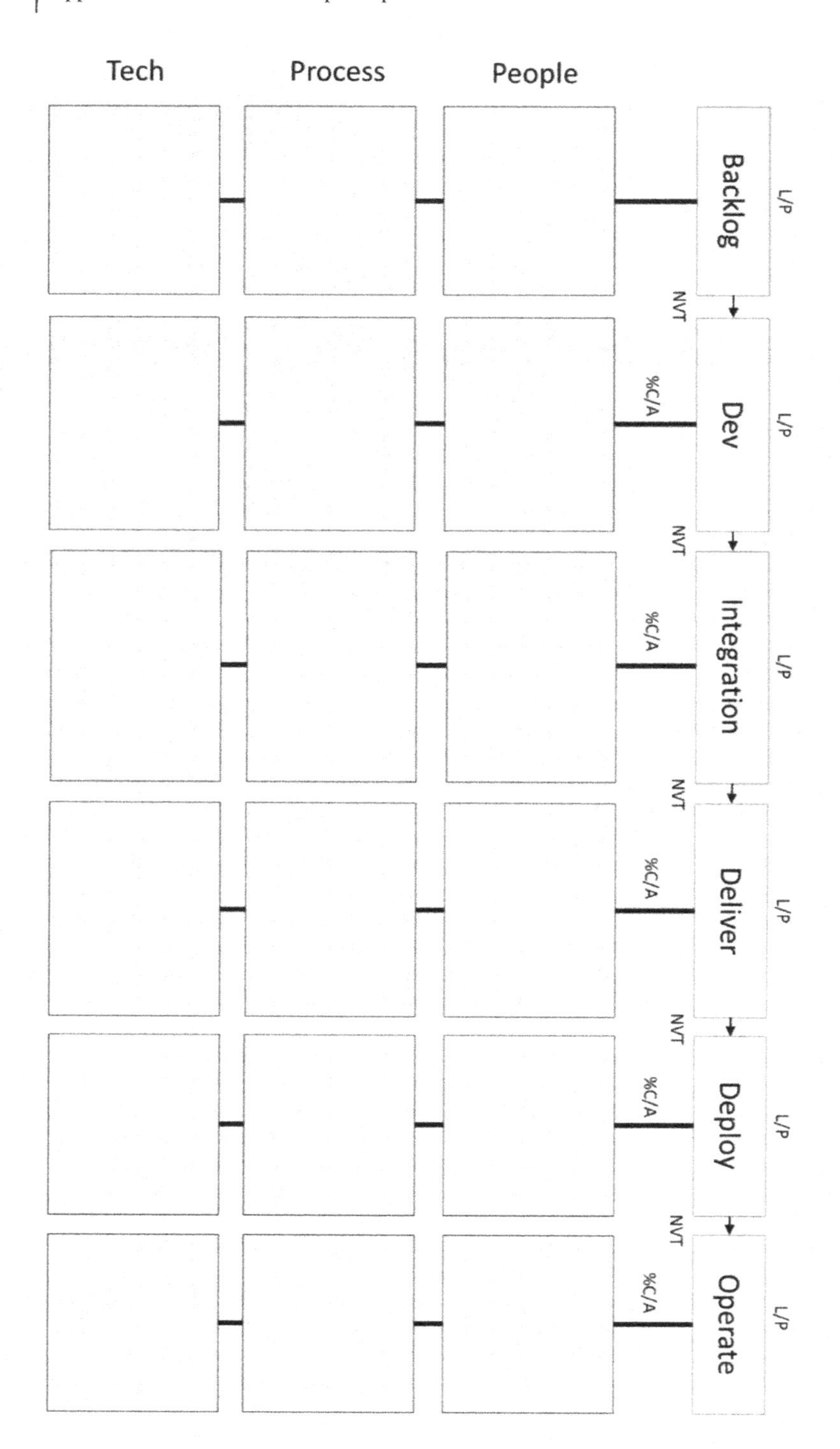

Tech
Process
People
Backlog
L/P
NVT
Dev
%C/A
L/P
NVT
Integration
%C/A
L/P
NVT
Deliver
%C/A
L/P
NVT
Deploy
%C/A
L/P
NVT
Operate
%C/A
L/P

Appendix L

DevOps Tools and Comparison Charts

The template on the following page is available in PowerPoint format on:

www.engineeringDevOps.com

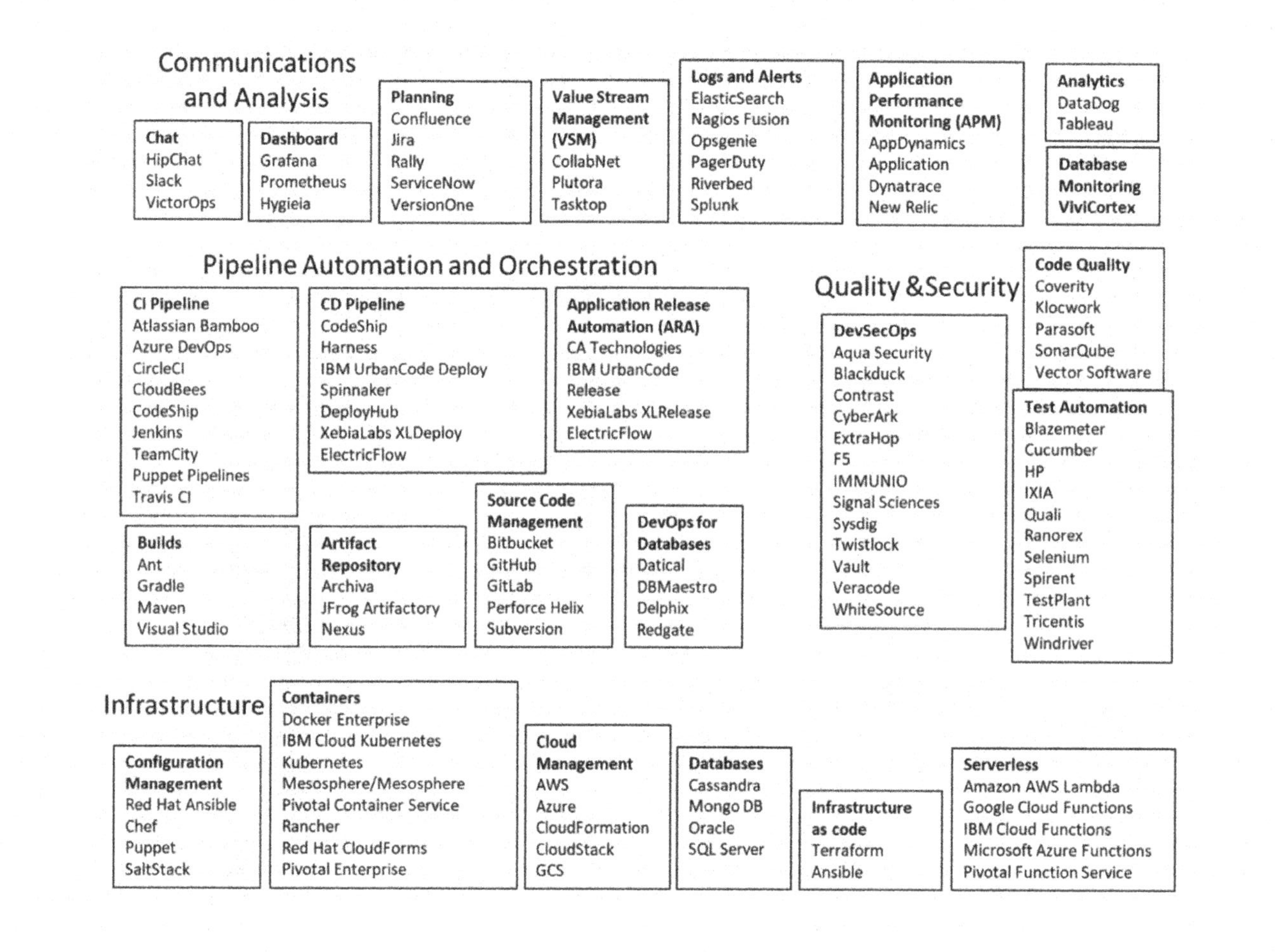
Communications and Analysis
Chat
HipChat
Slack
VictorOps
Dashboard
Grafana
Prometheus
Hygieia
Planning
Confluence
Jira
Rally
ServiceNow
VersionOne
Value Stream Management (VSM)
CollabNet
Plutora
Tasktop
Logs and Alerts
ElasticSearch
Nagios Fusion
Opsgenie
PagerDuty
Riverbed
Splunk
Application Performance Monitoring (APM)
AppDynamics
Application
Dynatrace
New Relic
Analytics
DataDog
Tableau
Database Monitoring
ViviCortex
Pipeline Automation and Orchestration
CI Pipeline
Atlassian Bamboo
Azure DevOps
CircleCI
CloudBees
CodeShip
Jenkins
TeamCity
Puppet Pipelines
Travis CI
CD Pipeline
CodeShip
Harness
IBM UrbanCode Deploy
Spinnaker
DeployHub
XebiaLabs XLDeploy
ElectricFlow
Application Release Automation (ARA)
CA Technologies
IBM UrbanCode Release
XebiaLabs XLRelease
ElectricFlow
Builds
Ant
Gradle
Maven
Visual Studio
Artifact Repository
Archiva
JFrog Artifactory
Nexus
Source Code Management
Bitbucket
GitHub
GitLab
Perforce Helix
Subversion
DevOps for Databases
Datical
DBMaestro
Delphix
Redgate
Quality &Security
Code Quality
Coverity
Klocwork
Parasoft
SonarQube
Vector Software
DevSecOps
Aqua Security
Blackduck
Contrast
CyberArk
ExtraHop
F5
IMMUNIO
Signal Sciences
Sysdig
Twistlock
Vault
Veracode
WhiteSource
Test Automation
Blazemeter
Cucumber
HP
IXIA
Quali
Ranorex
Selenium
Spirent
TestPlant
Tricentis
Windriver
Infrastructure
Configuration Management
Red Hat Ansible
Chef
Puppet
SaltStack
Containers
Docker Enterprise
IBM Cloud Kubernetes
Kubernetes
Mesosphere/Mesosphere
Pivotal Container Service
Rancher
Red Hat CloudForms
Pivotal Enterprise
Cloud Management
AWS
Azure
CloudFormation
CloudStack
GCS
Databases
Cassandra
Mongo DB
Oracle
SQL Server
Infrastructure as code
Terraform
Ansible
Serverless
Amazon AWS Lambda
Google Cloud Functions
IBM Cloud Functions
Microsoft Azure Functions
Pivotal Function Service

			Tool Alternative 1		Tool Alternative 2		Tool Alternative 3	
Comparison Category	Requirement	Weight (W) (1-5)	Description	Score (S) (1-5)	Description	Score (S) (1-5)	Description	Score (S) (1-5)
Source: Open-Source, Freemium, DIY, Enterprise								
Initial Cost: the licensing and labour cost for an initial configuration including costs internal to the organization.								
Total Cost of Ownership: the licensing and labour cost to the organization at scale, including costs internal to the organization.								
Compatibility: operating systems, ecosystems, cloud-native, DevOps frameworks, APIs								
Ease of Use: intuitive user interface, efficient controls, efficient outputs								
Administration Capabilities: installation support, diagnostic support, build-in metrics to track performance								
Functional Requirements: features and								
Non-Functional Requirements: Performance, reliability, stability, scalability								
Roadmap: planned enhancements for future solution requirements								
Support: Professional services for installation, proof-of-concept, configuration, training, and								
Weighted Score W x S								

Appendix M

Engineering DevOps Transformation RoadMap Template

The template on the following page is available in PowerPoint format on:

www.engineeringDevOps.com

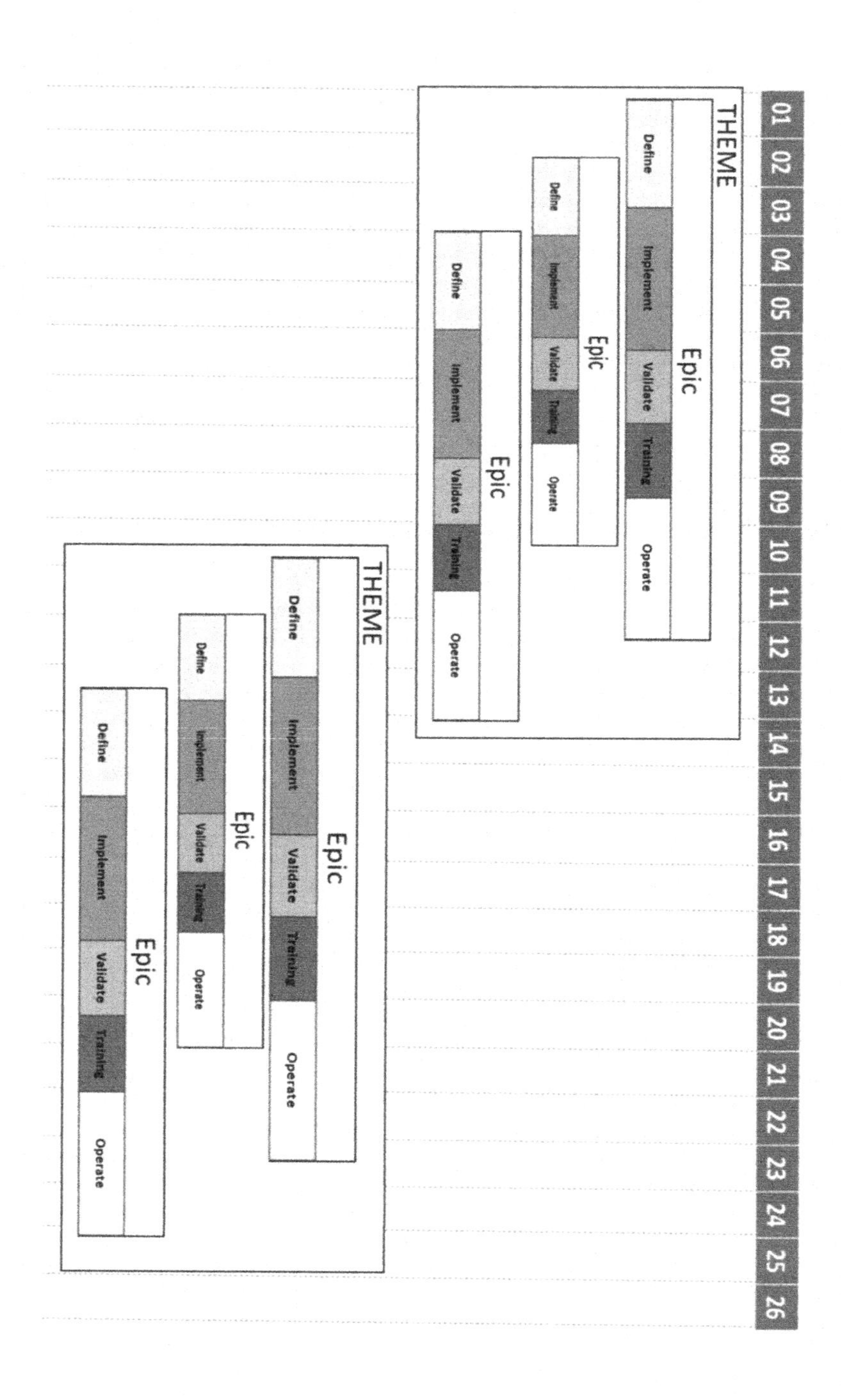
01
02
03
04
05
06
07
08
09
10
11
12
13
14
15
16
17
18
19
20
21
22
23
24
25
26
THEME
Epic
Define
Implement
Validate
Training
Operate
Epic
Define
Implement
Validate
Training
Operate
Epic
Define
Implement
Validate
Training
Operate
THEME
Epic
Define
Implement
Validate
Training
Operate
Epic
Define
Implement
Validate
Training
Operate
Epic
Define
Implement
Validate
Training
Operate

Appendix N

Engineering DevOps Transformation Backlog Template

This template is available in PowerPoint format on:

www.engineeringDevOps.com

EPIC	Phase	User Story	Task	Responsible
<theme a>	Define	<user story 1>	<task1>	<name>
<theme a>	Define	<user story 1>	<task2>	<name>
<theme a>	Define	<user story 1>	<task3>	<name>
<theme a>	Define	<user story 2>	<task1>	<name>
<theme a>	Define	<user story 2>	<task2>	<name>
<theme a>	Define	<user story 2>	<task3>	<name>
<theme a>	Define	<user story 3>	<task1>	<name>
<theme a>	Define	<user story 3>	<task2>	<name>
<theme a>	Define	<user story 3>	<task3>	<name>
<theme a>	Implement	<user story 1>	<task1>	<name>
<theme a>	Implement	<user story 1>	<task2>	<name>
<theme a>	Implement	<user story 1>	<task3>	<name>
<theme a>	Implement	<user story 2>	<task1>	<name>
<theme a>	Implement	<user story 2>	<task2>	<name>
<theme a>	Implement	<user story 2>	<task3>	<name>
<theme a>	Implement	<user story 3>	<task1>	<name>
<theme a>	Implement	<user story 3>	<task2>	<name>
<theme a>	Verify	<user story 1>	<task1>	<name>
<theme a>	Verify	<user story 2>	<task1>	<name>
<theme a>	Training	<user story 1>	<task1>	<name>
<theme a>	Training	<user story 2>	<task1>	<name>
<theme a>	Operate	<user story 1>	<task1>	<name>
<theme a>	Operate	<user story 2>	<task1>	<name>

Appendix O

Engineering DevOps Transformation ROI Calculator

The template on the following page is available in PowerPoint format on:

www.engineeringDevOps.com

Per Application (3 years Period)	Formula	NO DevOps Enhancements (n)	DevOps Enhancements (d)
R) Average # releases/year	**Estimate Number**		
A) Total Costs (3 years) $K	**B+C**		
B) Cost for 3 years of releases $K	**B1 + B4**		
B1) Labor Costs $K	3 x B2 x B3		
B2) Average Labor rate $K/year	Estimate Number		
B3) Average # workers	Estimate Number		
B4) Capital Depreciation (3 years amortization) $K	Estimate Number		
C) DevOps Enhancements costs $K	**C1 + C4**		
C1) Labor Costs$K	**3 x C2 x C3**		
C2) Average Labor rate $/year	**Estimate Number**		
C3) # workers for DevOps enhancement	**Estimate Number**		
C4) Capital Depreciation (3 years amortization) $K	**Estimate Number**		
D) Direct Savings Attributed to DevOps Enhancements	**An-Ad**		
F) Cost per release	**A/R**		
H) # Releases over 3 years	3 x R		
I) Additional releases due to DevOps Enhancements	**Hd - Hn**		
J) Costs of Equivalent # releases with DevOps enhancements (3 years) $K	**F x Hd**		
K) Equivalent Savings due to DevOps Enhancements (3 years) $K	**Jd - Jn**		
L) Return on Investment (ROI)	**K / C**		
M) Payback period (Months)	**C / (K/36)**		
E) Number of Applications	**Estimate Number**		
O) Direct Savings for all applications (3 years) $K	**E x Dd**		
P) Equivalent Savings for all applications (3 years) $K	**E x Kd**		

Appendix P

DevOps Transformation Solution Recommendation Meeting

This is an example presentation. The template is available in PowerPoint format on:

www.engineeringDevOps.com

DevOps Transformation Solution Recommendation Meeting

Attendees
Business Leaders
Finance
DevOps Sponsor
DevOps Initiator
DevOps Expert

Agenda

- DevOps Transformation Goals
- Summary of Current State
- Summary of Solution Requirements
- Future-State Value-Stream Map
- DevOps Transformation RoadMap
- Return-On-Investment
- Solution Recommendation Summary
- Next Steps

Strategic Goals for the DevOps Transformation

Our product organization shall be the market leader, measured by volume of sales, within three years, through increasing new product release velocity ten times while reducing non-value added costs in the delivery pipeline two times.

WHY? The market for this product category is maturing. Only the top three competitors are expected to survive the next 5 years. Industry trends show the winning companies will the ones with the fastest, most innovative value-stream and lowest price.

Definition of DevOps

DevOps is the application of lean practices Continuous Flow, Continuous Feedback and Continuous Improvement to people, process and technology for the benefit of agility, stability, efficiency, quality, security and satisfaction.

DevOps_The_Gray

There is no industry standard definition of DevOps yet it is important for each organization to have a definition the team can accept and rally around. The above definition is suggested. Alternative definitions are also acceptable provided the DevOps Transformation Team accepts a definition for use in the organization.

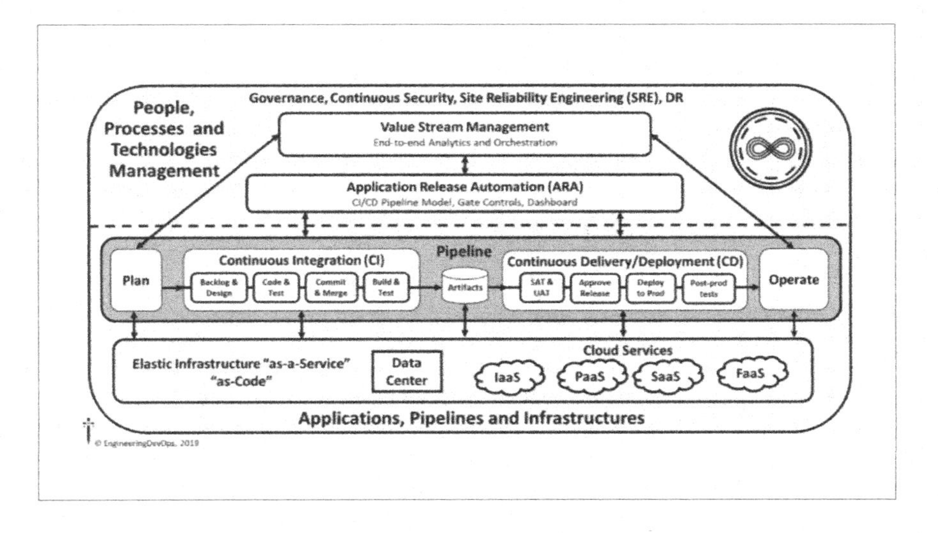

DevOps Transformation Application Selection

The DevOps transformation will apply to the following applications: <name of application(s)>

The <name> application will be used as a model for kicking-off and evaluating changes needed to people, process and technology for the DevOps transformation.

WHY? A DevOps Application Selection Scorecard indicates these application(s) will most likely benefit from the DevOps transformation.

DevOps Transformation Goals Workshop

For each goal enter Unit, Importance (I), Current State, Desired State.

Once all goal data is entered calculate % Improvement, Score and Rank to determine the top goals.

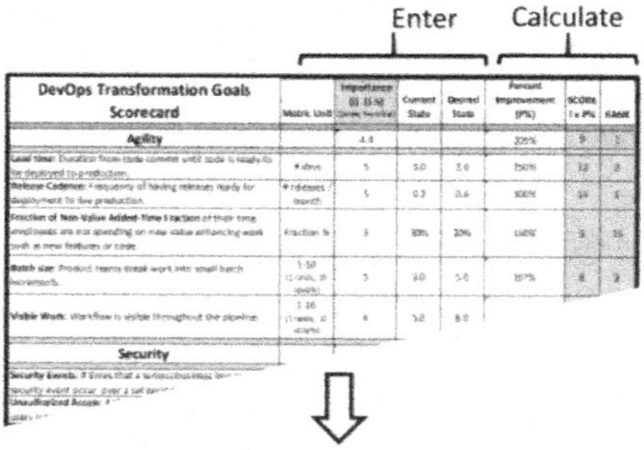

Enter

Calculate

DevOps Transformation Goals Scorecard

Agility

Security

20 goals in six categories: Agility, Stability, Efficiency, Quality, Security, and Satisfaction

DevOps Transformation Practices Score Workshop

For each Practice Category enter Importance and Current Level of Practice Score

Once all data is entered calculate a score and rank to determine the top priority practices.

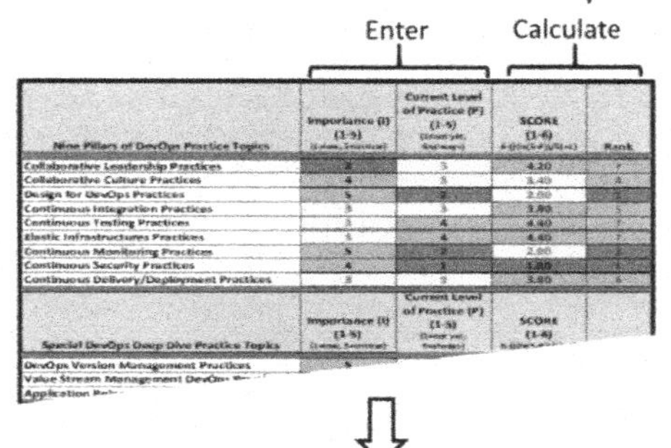

Nine Pillars and 11 Deep-Dive Topics

DevOps Solution Requirements

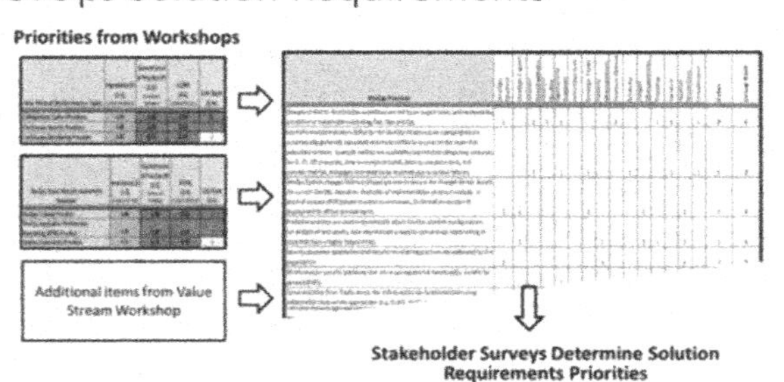

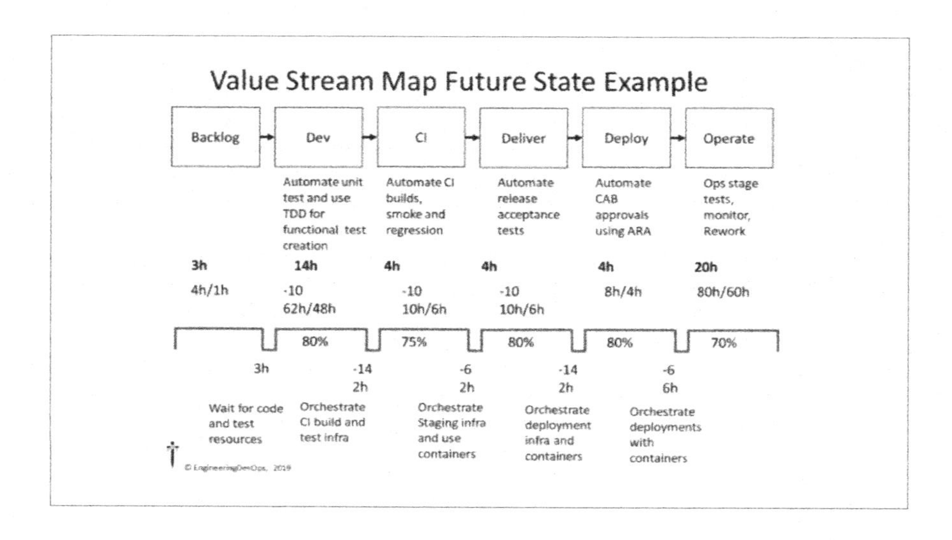
Value Stream Map Future State Example
Backlog
Dev
CI
Deliver
Deploy
Operate
Automate unit test and use TDD for functional test creation
Automate CI builds, smoke and regression
Automate release acceptance tests
Automate CAB approvals using ARA
Ops stage tests, monitor, Rework
3h
14h
4h
4h
4h
20h
4h/1h
-10
62h/48h
-10
10h/6h
-10
10h/6h
8h/4h
80h/60h
80%
75%
80%
80%
70%
3h
-14
2h
-6
2h
-14
2h
-6
6h
Wait for code and test resources
Orchestrate CI build and test infra
Orchestrate Staging infra and use containers
Orchestrate deployment infra and containers
Orchestrate deployments with containers

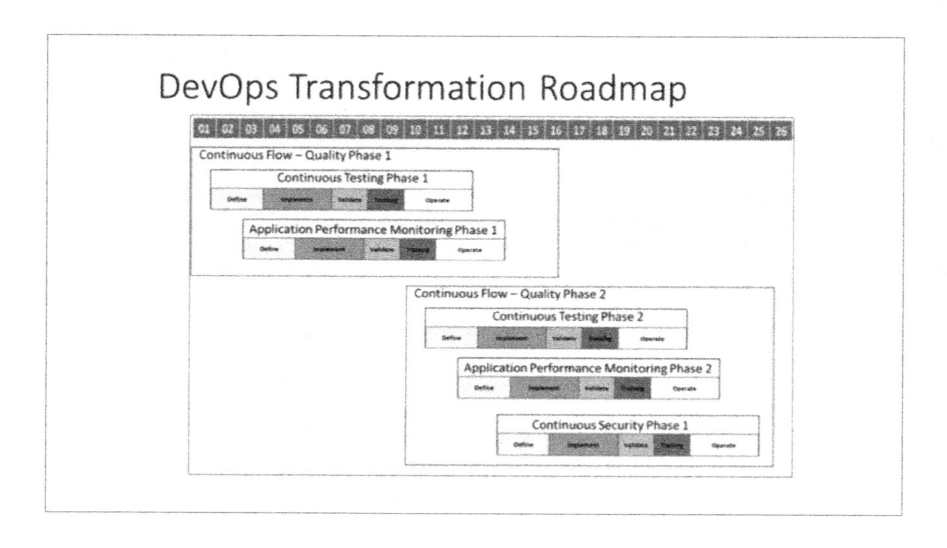
DevOps Transformation Roadmap
01 02 03 04 05 06 07 08 09 10 11 12 13 14 15 16 17 18 19 20 21 22 23 24 25 26
Continuous Flow – Quality Phase 1
Continuous Testing Phase 1
Application Performance Monitoring Phase 1
Continuous Flow – Quality Phase 2
Continuous Testing Phase 2
Application Performance Monitoring Phase 2
Continuous Security Phase 1

DevOps Transformation Return On Investment

Theme	Epic	Costs $K 3 years	Savings $K 3 years	ROI Ratio	Payback Months
DevOps Phase 1	Continuous Testing 1	500	7,500	15	7
	Monitoring 1	350	6,300	18	7
DevOps Phase 2	Continuous Testing 2	300	6,000	20	5
	Monitoring 2	250	3,750	15	7
	Security 1	400	8,800	22	8
TOTAL	ALL	1,800	32,350	18	7

Assumptions:
3 Years amortization
8 Applications

Primary source of Savings:
1) Automation of test phases
2) Bottlenecks removed in the value stream provided by improved visibility
3) Reduction of Non-Value-Added work

Risks if not approved:
1) 18 releases will be delayed.
2) Security risks

DevOps Solution Recommendation Summary

- The DevOps Solution recommended meets the strategic goals
- ROI and payback period is sufficient to justify the projects

Approval is requested for the DevOps Solution Recommendation

Next Step Actions towards Realizing the DevOps Transformation

Action	Date	Responsible
Schedule date to reply to actions	MM/DD/YYYY	Initiator
Schedule dates for approvals of Epics	MM/DD/YYYY	Initiator
Assign Backlog	MM/DD/YYYY	Initiator
Set dates for follow-up with business leaders	MM/DD/YYYY	Initiator

Appendix Q

NetDevOps Blueprint

The blueprint can be found on the following page.

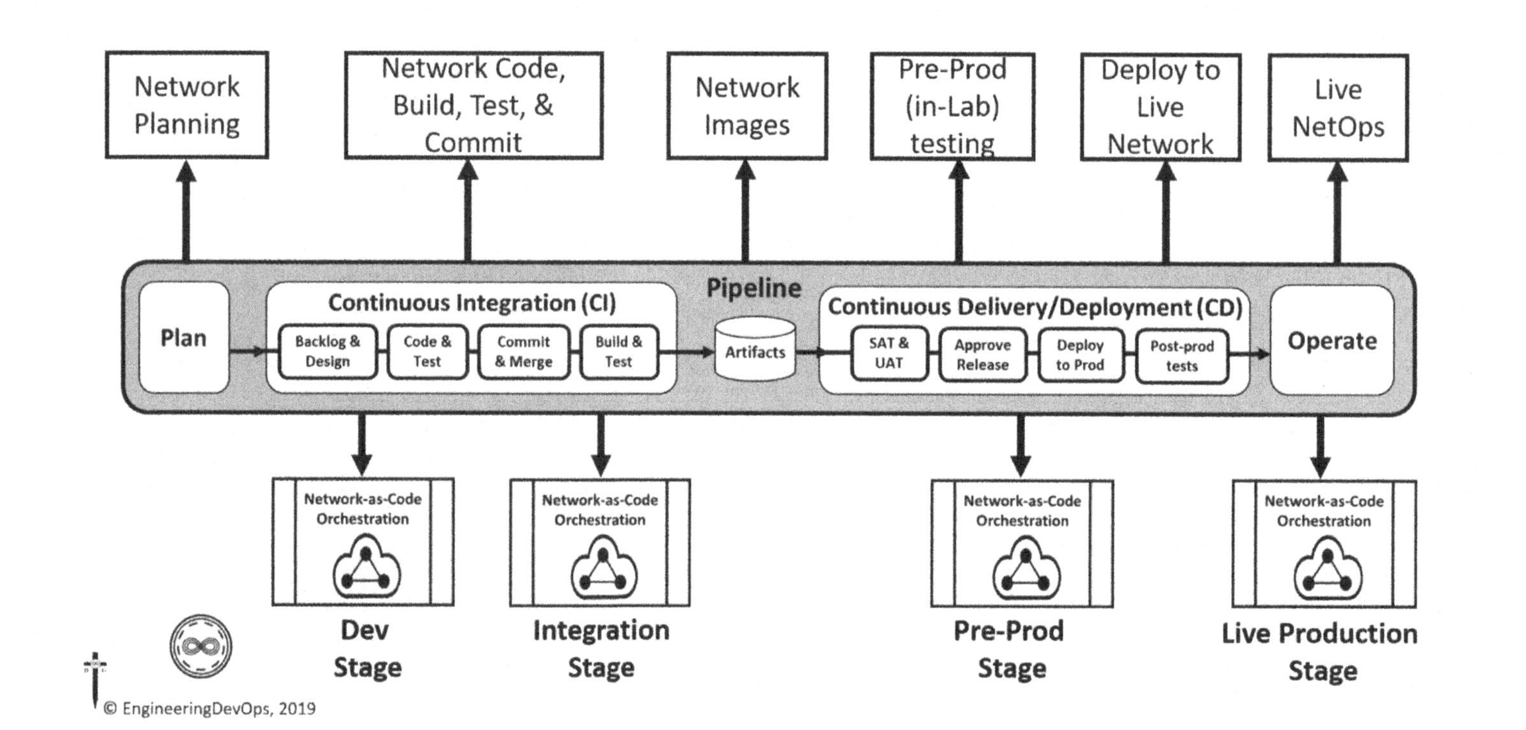
Network Planning
Network Code, Build, Test, & Commit
Network Images
Pre-Prod (in-Lab) testing
Deploy to Live Network
Live NetOps
Pipeline
Plan
Continuous Integration (CI)
Backlog & Design
Code & Test
Commit & Merge
Build & Test
Artifacts
Continuous Delivery/Deployment (CD)
SAT & UAT
Approve Release
Deploy to Prod
Post-prod tests
Operate
Network-as-Code Orchestration
Dev Stage
Network-as-Code Orchestration
Integration Stage
Network-as-Code Orchestration
Pre-Prod Stage
Network-as-Code Orchestration
Live Production Stage
© EngineeringDevOps, 2019

References

Book references (B)

B1. W. Edwards Deming, *Out of the Crisis*, MIT Press, First Edition, 1982.

B2. Eliyahu M. Goldratt, *The Goal*, North River Press, First Edition, 1984.

B3. Gene Kim, Kevin Behr, and George Spafford, *The Phoenix Project*, IT Revolution Press, First Edition, 2013.

B4. Glenford Myers, *The Art of Software Testing*, First Edition, 1979.

B5. Kent Beck, *Extreme Programming Explained*, First Edition, 1999.

B6. Jez Humble and David Farley, *Continuous Delivery*, Addison Wesley, First Edition, 2011.

B7. Gene Kim, Jez Humble, Patrick Debois, and John Willis, *The DevOps Handbook*, IT Revolution Press, First Edition, 2016.

B8. Nicole Forsgren, Ph.D., Jez Humble and Gene Kim, *Accelerate*, IT Revolution Press, First Edition, 2017.

B9. Jennifer Davis and Katherine Daniels, *Effective DevOps*, O'Reilly, First Edition, 2016, p3, pp1.

B10. Jennifer Davis and Katherine Daniels, *Effective DevOps*, O'Reilly, First Edition, 2016, p7, pp1.

B11. Karen Martin and Mike Osterling, *Value-Stream Mapping*, McGraw Hill, First Edition, 2014.

B12. Gary Gruver, Mike Young, and Pat Fulghum, *A Practical Approach to Large-Scale Agile Development: How HP Transformed LaserJet FutureSmart Firmware*, Addison-Wesley, First Edition, 2013.

B13. Betsy Beyer, Chris Jones, Jennifer Petoff, and Niall Richard Murphy, *Site Reliability Engineering*, O'Reilly, 2016, p7.

B14. Edward Wolff, *Microservices Flexible Software Architectures*, Leanpub, 2016.

B15. Newman, *Building Microservices*, O'Reilly, 2015.

B16. Len Bass, Ingo Weber, and Liming Zhu, *DevOps: A Software Architect's Perspective*, Addison Wesley, 2015.

B17. Josh Stella, *Immutable Infrastructure*, O'Reilly, 2015.

B18. Michael J. Kavis, *Architecting the Cloud*, Wiley, 2014.

B19. Jeff Geerling, *Ansible for DevOps*, Leanpub, 2019.

B20. Sanjeev Sharma, *The DevOps Adoption Playbook*, Wiley and IBM Press, 2017.

B21. Michael S. Cuppett, *DevOps DBAs, and DBaaS*, Apress, 2016.

B22. Kelsey Hightower, *Kubernetes Up & Running*, O'Reilly, 2017.

B23. George V. Hulme, *DevSecOps*, DevOps.com, 2018.

B24. Dr. Bhaskar Ghosh, *DevOps for the Modern Enterprise*, IT Revolution Press., 2018.

Courses (C)

C1. www.devopsinstitute.com

C2. www.udemy.com

Meeting and Events (E)

E1. https://events.itrevolution.com/

Research and Industry Reports (R)

R1. Puppet Labs and DORA, "State of DevOps Report," Issues 2016, 2017, 2018.

R2. Puppet Labs, "State of DevOps Market Segmentation Report," 2017.
R3. Capgemini, Sogeti, and Micro Focus, "World Quality Report," Issues 2016–2017, 2017–2018.
R4. W.S. Humphrey, March 1988. "Characterizing the Software Process: A Maturity Framework". IEEE Software 5(2):73–79. doi:10.1109/52.2014. ISSN 0740-7459.
R5. C.H. Lee, M.W.A. Hornbeek, and T. Naujokas. "A Network Test Environment for Packet Switched Networks," IEE&C proceedings, 1983.
R6. C.H. Lee, M.W.A. Hornbeek, and C. Duong, "Network Software Engineering Capability and Its Delivery," ISS'84, May 1984
R7. M.W.A. Hornbeek, "An Integrated Test Center For SL-10 Packet Networks," ACM proceedings, 1985.
R8. M.W.A. Hornbeek, "TEAM: A Desktop Window to the Captive Office," Telesis #1, 1989.
R9. Gartner, "Magic Quadrant for Application Release Automation."
R10. Forester, "Continuous Delivery and Release Automation."
R11. Forester, "Elevate Agile-Plus-DevOps with Value-Stream Management."
R12. George V. Hulme, "The State of Security," RSA Special Report.
R13. "Software Engineering 2014 Curriculum Guidelines for Undergraduate Degree Programs in Software Engineering," 23 February 2015

Standards References (S)

S1. IEEE Standard Glossary of Software Engineering Terminology, IEEE std 610.12-1990, 1990.
S2. John H. Baumert and Mark S. McWhinney, "Software Measures and the Capability Maturity Model," September 1992.
S3. https://www.itsmacademy.com/
S4. https://searchcloudcomputing.techtarget.com/definition/TOSCA-Topology-and-Orchestration-Specification-for-Cloud-Applications

Trademarks (T)

T1. CMM is a registered service mark of Carnegie Mellon University (CMU).

Web References (W)

W1. https://devops.com/defining-devops-how-do-you-know-when-you-have-achieved-devops/
W2. https://devops.com/the-origins-of-devops-whats-in-a-name/
W3. https://agilemanifesto.org/
W4. https://www.scaledagileframework.com/about/
W5. https://puppet.com/docs
W6. https://devops-research.com/
W7. https://dougseven.com/2014/04/17/knightmare-a-devops-cautionary-tale/
W8. https://www.computer.org/
W9. https://services.acm.org
W10. http://peo.on.ca/index.php/ci_id/21643/la_id/1.htm
W11. https://www.acm.org/binaries/content/assets/education/se2014.pdf Page 15 paragraph 2
W12. https://www.helpnetsecurity.com/2018/05/22/open-source-code-security-risk/
W13. https://opensourceforu.com/2015/04/open-source-software-engineering-an-introduction-to-open-source-tools/
W14. https://www.sei.cmu.edu/
W15. https://cmmiinstitute.com/
W16. https://en.wikipedia.org/wiki/Conway%27s_law
W17. https://devops.com/using-calms-to-assess-organizations-devops/
W18. https://www.asme.org/engineering-topics/articles/manufacturing-design/5-lean-principles-every-should-know
W19. https://en.wikipedia.org/wiki/Six_Days,_Seven_Nights
W20. https://devops.com/design-devops-best-practices/
W21. https://12factor.net/

W22. https://en.wikipedia.org/wiki/Field_of_Dreams
W23. https://www.techrepublic.com/article/10-steps-to-devops-success-in-the-enterprise/
W24. https://infocus.dellemc.com/bart_driscoll/can-devops-continuous-delivery-work-commercial-off-shelf-software/
W25. https://electric-cloud.com/blog/5-aspects-that-makes-continuous-delivery-for-embedded-different/
W26. https://www.businesswire.com/news/home/20030820005031/en/EdenTree-Technologies-Launches-Infrastructure-Automation-Solution-Equipment
W27. https://www.itworldcanada.com/article/devops-does-not-negate-itil-or-itsm-they-can-be-leveraged-for-greater-agility/379603
W28. https://techbeacon.com/enterprise-it/4-ways-marry-itsm-devops
W29. https://www.forbes.com/sites/janakirammsv/2019/02/27/how-ready-are-your-applications-for-the-cloud-a-look-at-the-enterprise-application-maturity-model/#15f283e760f9
W30. https://www.oreilly.com/ideas/modules-vs-microservices
W31. https://en.wikipedia.org/wiki/Service-oriented_architecture
W32. https://www.archives.gov/preservation/products/definitions/products-services.html
W33. https://en.wikipedia.org/wiki/Platform_as_a_service
W34. https://en.wikipedia.org/wiki/Infrastructure_as_a_service
W35. https://en.wikipedia.org/wiki/Idempotence
W36. https://searchitoperations.techtarget.com/definition/immutable-infrastructure
W37. https://en.wikipedia.org/wiki/Serverless_computing
W38. https://en.wikipedia.org/wiki/Function_as_a_service
W39. https://impaddo.com/blog/serverless-architecture/
W40. https://www.gartner.com/binaries/content/assets/events/keywords/applications/apps20i/retire_the_threetier_applica_308298.pdf
W41. https://kubernetes.io/docs/concepts/workloads/controllers/deployment/
W42. https://webinars.devops.com/the-state-of-devops-tools-a-primer
W43. https://xebialabs.com/periodic-table-of-devops-tools/

W44. https://www.leanblog.org/2011/05/guest-post-what-is-yokoten/
W45. https://devops.com/devoptimism-the-future-of-devops-is-becoming-clearer/
W46. https://devops.com/6-ways-ai-and-ml-will-change-devops-for-the-better/
W47. https://techbeacon.com/enterprise-it/6-open-source-trends-will-shape-it-ops
W48. https://www.rtinsights.com/iot-and-devops-security-integration-deployment/
W49. https://www.cloudcomputing-news.net/news/2019/jan/14/three-key-predictions-cloud-industry-2019-multi-cloud-governance-and-blurred-lines/
W50. https://logz.io/blog/nanoservices-vs-microservices/
W51. https://www.slideshare.net/JulesPierreLouis/from-monolith-to-microservices-and-beyond
W52. https://youtu.be/3tKM2MF9ilg
W53. https://devops.com/how-devops-can-create-smarter-more-agile-5g-telecom-networks/
W54. http://trace3.com/blog/7-pillars-of-devops-essential-foundations-for-enterprise-success/
W55. https://containerjournal.com/2018/10/16/9-pillars-of-containers-best-practices/
W56. https://devops.com/9-pillars-of-continuous-security-best-practices/
W57. http://trace3.com/blog/can-culture-detract-from-success-with-devops/
W58. http://devops.com/blogs/devops-continuous-testing/
W59. http://devops.com/blogs/continuous-testing-accelerated/
W60. http://devops.com/blogs/qa-continuous-testing/
W61. https://devops.com/unnatural-devops-delivers-supernatural-results/
W62. https://devops.com/dev-vs-ops-needs-large-scale-version-control-systems/
W63. https://webinars.devops.com/why-value-stream-management-is-essential-for-effective-devops

W64. https://devops.com/nothing-less-total-devops-will-guarantee-results/

W65. https://www.upwork.com/hiring/for-clients/aws-vs-azure-vs-google-cloud-platform-comparison/

W66. https://www.cloudhealthtech.com/blog/aws-vs-azure-vs-google

W67. https://devops.com/12-tips-managing-multi-cloud-environment/

W68. https://www.slideshare.net/MarcHornbeek/devops-evolution-the-next-generation?qid=2e67e173-cf28-450a-ac44-53907d44d5fa&v=&b=&from_search=10

W69. https://appdevelopermagazine.com/2993/2015/7/21/DevOps-Continuous-Testing-Requires-Speed-with-Relevance/

W70. http://computer.ieee-bv.org/wp-content/uploads/2015/10/2015-09-09-DevOps.pdf

W71. http://www.slideshare.net/Spirent/ieee-buenaventura-cs-chapter-march-9-2016-v4

W72. http://www.perforce.com/sites/default/files/continous-change-driven-build-verification-wp.pdf

W73. https://infocus.dellemc.com/bart_driscoll/can-devops-continous-delivery-work-commercial-off-shelf-software/

W74. https://en.wikipedia.org/wiki/Chaos_engineering

W75. https://www.owasp.org/index.php/Category:OWASP_Top_Ten_Project

About the Author

Marc Hornbeek is known as DevOps_the_Gray Esq. DevOps is his career passion. He is a senior member of the Institute of Electrical and Electronics Engineers (IEEE) with 43 years of membership and is qualified as a registered professional engineer in Ontario, Canada. He was awarded "Outstanding Engineer of 2016" by IEEE Region 6 (Western USA) for lifetime work on "automation applied to development and testing of networks, systems, protocols, labs and DevOps" (https://ieee-region6.org/2016/2016-region-6-awards-winners/).

Marc's approach to DevOps is as an engineering leader and consultant. He has led over sixty DevOps transformation projects for a variety of applications and organizations including large enterprises; wireless, mobile, and IOT product manufacturers; service providers; COTS systems; and government agencies. He's worked with AAA, Cisco, DaVita, ECI, Ericsson, Herbalife, GSI, LDS, NetApp, Nokia, PWCgov, Spirent, Tekelec, and WAM. Marc developed the concept ***"Nine Pillars of DevOps,"*** created ***"DevOps Engineering Blueprints,"*** compiled an extensive database of ***"Recommended Engineering Practices,"*** and invented the ***"DevOps Seven-Step Transformation Engineering Blueprint."*** These DevOps engineering tools, described in this book, provide prescriptive reference guides for leaders and practitioners of DevOps to calibrate their organization's DevOps vision, get alignment, perform assessments, create solutions, realize DevOps implementations, operationalize, and expand DevOps successfully across their organization.

As a Principal Consultant of DevOps at Trace3 and EngineeringDevOps, Marc has lead DevOps assessments and strategic planning for *Fortune* 2000 clients across the United States. This involves working with leaders and key practitioners that need a practical roadmap to guide their DevOps transformation.

As an affiliate of the DevOps Institute (www.devopsinstitute.com), Marc authored the Continuous Delivery Architect (CDA) and the DevOps Test Engineering (DTE) certification courses, which are offered by affiliated global training partners. As a freelance consultant and author, Marc writes e-books, webinars, and white papers for clients. He is a blogger on DevOps.com (www.devops.com), as well as other sites. His writings cover a broad range of DevOps and engineering topics. He regularly speaks at DevOps industry events, webinars, and meetups.

Marc's degrees and certifications include an Executive MBA—Pepperdine University, California; an Engineering B.Sc. (Honors)—Queen's University, Canada; and the following DevOps certifications from the DevOps Institute: DevOps Foundations, DevOps Test Engineering (DTE), Continuous Delivery Architect (CDA), and DevOps Leader.

Marc's hobbies (besides DevOps!) include sailing, woodworking, maritime history, Elizabethan history and Arthurian facts and fiction, and frequent travel. He resides in his home ***Casa Avalon***, near Puerto Vallarta, Mexico, with his son Michael. You can connect with Marc via LinkedIn or through his personal website.

Marc Hornbeek, DevOps_the_Gray Esq.
https://www.linkedin.com/in/marchornbeek
www.EngineeringDevOps.com